Communications
in Computer and Information Science   893

Commenced Publication in 2007
Founding and Former Series Editors:
Phoebe Chen, Alfredo Cuzzocrea, Xiaoyong Du, Orhun Kara, Ting Liu,
Dominik Ślęzak, and Xiaokang Yang

More information about this series at http://www.springer.com/series/7899

Elias Pimenidis · Chrisina Jayne (Eds.)

Engineering Applications of Neural Networks

19th International Conference, EANN 2018
Bristol, UK, September 3–5, 2018
Proceedings

 Springer

Editors
Elias Pimenidis 🆔
University of the West of England
Bristol
UK

Chrisina Jayne 🆔
Oxford Brookes University
Oxford
UK

ISSN 1865-0929   ISSN 1865-0937   (electronic)
Communications in Computer and Information Science
ISBN 978-3-319-98203-8   ISBN 978-3-319-98204-5   (eBook)
https://doi.org/10.1007/978-3-319-98204-5

Library of Congress Control Number: 2018950649

This Springer imprint is published by the registered company Springer Nature Switzerland AG
The registered company address is: Gewerbestrasse 11, 6330 Cham, Switzerland

Preface

This CCIS Springer volume contains the proceedings of the 19th International Conference on Engineering Applications of Neural Networks (EANN 2018), which was held at the University of the West of England, Bristol, UK, during September 3–5, 2018. The conference was supported by the department of Computer Science and Creative Technologies and the Faculty of Environment and Technology of the University of the West of England.

With a history of 23 years since the inaugural event in Otaniemi, Finland, in 1995, EANN has grown in reputation and has been attracting established as well as new researchers in the field of neural networks and other related areas of artificial intelligence. Since 2009, EANN has be supported technically by the International Neural Network Society (INNS).

EANN 2018 attracted submissions from countries around the globe with the presenting authors/delegates coming from 12 countries, namely, Belgium, China, Germany, Italy, Japan, Norway, Oman, Pakistan, Portugal, Turkey, the USA, and the UK. The volume includes 16 papers that were accepted for oral presentation as full papers (at 12 pages each), representing a 41% acceptance rate on the number of submissions. The Program Committee encouraged by the quality of submissions by young researchers also accepted a further five submissions to be presented as short papers and are included in this volume. All papers were subject to a rigorous peer-review process by at least two independent academic referees, members of the international Program Committee.

The accepted papers demonstrate a variety of novel neural network and other artificial intelligence approaches applied to challenging real-world problems. The papers cover topics such as: activity recognition, deep learning, extreme learning machines, fuzzy systems, machine learning applications, predictive models, recommender systems, recurrent neural networks, and spiking neural networks applications.

The following keynote speakers were invited and gave lectures on exciting neural network application topics:

- Professor Plamen P. Angelov, Director of LIRA (Lancaster Intelligent, Robotic and Autonomous systems) Research Centre, Lancaster University, UK
- Professor Anthony Piper, Deputy Director of the Bristol Robotics Laboratory, UWE, Bristol, UK
- Professor Chrisina Jayne, Head of School of Engineering, Computing and Mathematics, Oxford Brookes University, UK

On behalf of the conference Organizing Committee, we would like to thank all those who contributed to the organization of this year's program, and in particular the Program Committee members.

September 2018

Elias Pimenidis
Chrisina Jayne

Organization

General Chairs

Anthony Pipe	University of the West of England, UK
Elias Pimenidis	University of the West of England, UK

Organizing Chairs

Steve Battle	University of the West of England, UK
Antisthenis Tsompanas	University of the West of England, UK

Program Chairs

Chrisina Jayne	Oxford Brookes University, UK
Phil Legg	University of the West of England, UK

Advisory Chairs

Ilias Maglogiannis	University of Piraeus, Greece
Lazaros Iliadis	Democritus University of Thrace, Greece

Workshop Chairs

Mehmet Aydin	University of the West of England, UK
Nikolaos Polatidis	University of Brighton, UK

Publicity Chair

Plamen Angelov	Lancaster University, UK

Honorary Committee

John MacIntyre	University of Sunderland, UK
Nikola Kasabov	Auckland University of Technology, New Zealand

Publications Chair

Antonios Papaleonidas	Democritus University of Thrace, Greece

Student Forum Chairs

Rong Yang   University of the West of England, UK
Emmanuel Ogunshile   University of the West of England, UK

Website Chair

Martyn Fitzgerald   University of West of England, UK

Program Committee

Michel Aldanondo	Toulouse University CGI, France
Athanasios Alexiou	Novel Global Community Educational Foundation, Australia
Ioannis Anagnostopoulos	University of Thessaly, Greece
Plamen Angelov	Lancaster University, UK
Mehmet Emin Aydin	University of the West of England, UK
Costin Badica	University of Craiova, Romania
Rashid Bakirov	Bournemouth University, UK
Zbigniew Banaszak	Warsaw University of Technology, Poland
Steve Battle	University of the West of England, UK
Bartlomiej Beliczynski	Warsaw University of Technology, Poland
Kostas Berberidis	University of Patras, Greece
Nik Bessis	Edge Hill University, UK
Monica Bianchini	Università di Siena, Italy
Giacomo Boracchi	Politecnico di Milano, Italy
Farah Bouakrif	University of Jijel, Algeria
Anne Canuto	Federal University of Rio Grande do Norte, Brazil
Diego Carrera	Politecnico di Milano, Italy
George Caridakis	National Technical University of Athens, Greece
Ioannis Chamodrakas	National and Kapodistrian University of Athens, Greece
Aristotelis Chatziioannou	National Hellenic Research Foundation, Greece
Ben Daubney	MBDA, UK
Jefferson Rodrigo De Souza	FACOM/UFU, Brazil
Kostantinos Demertzis	Democritus University of Thrace, Greece
Ioannis Dokas	Democritus University of Thrace, Greece
Ruggero Donida Labati	Politecnico di Milano, Italy
Mauro Gaggero	National Research Council of Italy, Italy
Christos Georgiadis	University of Macedonia, Greece
Giorgio Gnecco	IMT - Institute for Advanced Studies, Lucca, Italy
Denise Gorse	University College London, UK
Foteini Grivokostopoulou	University of Patras, Greece
Xiaowei Gu	Lancaster University, UK
Hakan Haberdar	University of Houston, USA
Petr Hajek	University of Pardubice, Czech Republic

Ioannis Hatzilygeroudis	University of Patras, Greece
Martin Holena	Institute of Computer Science, Czech Republic
Amin Hosseinian-Far	University of Northampton, UK
Chrisina Jayne	Oxford Brookes University, UK
Jacek Kabzinski	Lodz University of Technology, Poland
Antonios Kalampakas	Democritus University of Thrace, Greece
Ryotaro Kamimura	Tokai University, Japan
Dmitry Kangin	University of Exeter, UK
Stelios Kapetanakis	University of Brighton, UK
Ioannis Karydis	Ionian University, Greece
Petros Kefalas	The University of Sheffield, UK
Muhammad Khurram Khan	King Saud University, Saudi Arabia
Kyriaki Kitikidou	Democritus University of Thrace, Greece
Avrilia Kogketsof	Democritus University of Thrace, Greece
Yiannis Kokkinos	University of Macedonia, Greece
Mikko Kolehmainen	University of Eastern Finland, Finland
Petia Koprinkova-Hristova	Bulgarian Academy of Sciences, Bulgaria
Konstantinos Koutroumbas	National Observatory of Athens, Greece
Paul Krause	University of Surrey, UK
Ondrej Krejcar	University of Hradec Kralove, Czech Republic
Efthyvoulos Kyriacou	Frederick University, Cyprus
Phil Legg	University of the West of England, UK
Florin Leon	Gheorghe Asachi Technical University of Iasi, Romania
Aristidis Likas	University of Ioannina, Greece
Spiros Likothanassis	University of Patras, Greece
Ilias Maglogiannis	University of Piraeus, Greece
George Magoulas	Birkbeck College, UK
Mario Natalino Malcangi	Università degli Studi di Milano, Italy
Francesco Marcelloni	University of Pisa, Italy
Konstantinos Margaritis	University of Macedonia, Greece
Nikolaos Mitianoudis	Democritus University of Thrace, Greece
Valeri Mladenov	Technical University Sofia, Greece
Haralambos Mouratidis	University of Brighton, UK
Phivos Mylonas	National Technical University of Athens, Greece
Stavros Ntalampiras	Politecnico di Milano, Italy
Mihaela Oprea	University Petroleum-Gas of Ploiesti, Romania
Elpiniki Papageorgiou	Technological Educational Institute of Lamia, Greece
Basil Papadopoulos	Democritus University of Thrace, Greece
Antonios Papaleonidas	Democritus University of Thrace, Greece
Michalis Pavlidis	University of Brighton, UK
Daniel Pérez	University of Oviedo, Portugal
Isidoros Perikos	University of Patras, Greece
Miltos Petridis	Middlesex University London, UK
Elias Pimenidis	University of the West of England, UK
Anthony Pipe	University of the West of England, UK

Contents

Predictive Models, Fuzzy and Recommender Systems

Recurrent Neural Networks and Spiking Neural Networks

Activity Recognition

A Framework for Semi-Supervised Adaptive Learning for Activity Recognition in Healthcare Applications

Prankit Gupta[✉] and Praminda Caleb-Solly

Bristol Robotics Laboratory, University of the West of England, Bristol BS16 1QY, UK
prankit.gupta@brl.ac.uk, praminda.caleb-solly@uwe.ac.uk

Abstract. With the growing popularity of the Internet of Things and connected home products, potential healthcare applications in a smart-home context for assisted living are becoming increasingly apparent. However, challenges in performing real-time human activity recognition (HAR) from unlabelled data and adapting to changing user health remain a major barrier to the practicality of such applications. This paper aims to address these issues by proposing a semi-supervised adaptive HAR system which combines offline and online recognition techniques to provide intelligent real-time support for frequently repeated user activities. The viability of this approach is evaluated by pilot testing it on data from the Aruba CASAS dataset, and additional pilot data collected in the Bristol Robotics Lab's Assisted Living Studio. The results show that 71% of activity instances were discovered, with an F1-score of 0.93 for the repeating "Meal_Prep" activities. Furthermore, real-time recognition on the collected pilot data occurred near the beginning of the activity 64% of the time and at the halfway point in the activity 96% of the time.

Keywords: Activity recognition · Smart-home · Healthcare

1 Introduction

The term smart-home (SH) refers to the concept of integrating everyday items/home appliances with various sensors, actuators, and relays, and their inter-networking to achieve various forms of automation [1]. An increasing population of older adults with ageing-related impairments and long-term conditions is putting a greater amount of pressure on healthcare providers, who are struggling to balance high-quality care with reduced budgets. The utilization of IoT technology within a SM context to provide Ambient Assisted Living (AAL) is regarded as a key solution to achieve this balance [2].

An important area of research in SHs for AAL is human activity recognition (HAR). Offline and online HAR are two different approaches for detecting activities, with online methods working in real-time as the activity is occurring [3]. Most conventional HAR approaches that have shown promising results in the past have used supervised learning, requiring large amounts of expert-labelled user activity training data which is both difficult to obtain and not practical for real-world deployment [4]. These systems also

© Springer Nature Switzerland AG 2018
E. Pimenidis and C. Jayne (Eds.): EANN 2018, CCIS 893, pp. 3–15, 2018.
https://doi.org/10.1007/978-3-319-98204-5_1

largely remain unable to adapt to changing user health and behaviour over time, as they require supervised re-training with new activity data to account for this.

This paper proposes an alternative approach to conventional HAR, where unlike traditional systems the focus is not on training a classifier to identify every instance of every activity, but rather on discovering frequently repeating activity patterns. Accordingly, this paper presents a framework which utilizes a pre-defined short-term system memory for performing offline HAR on unlabelled data and updating the user activity model using Bayesian Networks for real-time detection. This combination of offline and online HAR works simultaneously to constantly learn activity patterns and evolve the user activity model on a short-term basis to account for any new or changing activities. This evolving user activity model could be used to set-up automated prompts for users with cognitive decline as well as track and continually assess user health to complement carer support. An implementation of this approach is also pilot tested with the Aruba CASAS dataset [5] and pilot data collected in the Assisted Living Studio (ALSt) which is set up as a realistic home environment in the Bristol Robotics Laboratory.

The rest of the paper is divided as follows – Sect. 2 provides background information and outlines related work; Sect. 3 formulates the problem; Sect. 4 presents the overall framework for the HAR system, including descriptions of the offline and online HAR systems and machine learning methods used; Sect. 5 describes the data sets used for pilot testing the approach; Sect. 6 outlines the procedure followed in developing and testing the system; Sect. 7 presents the results and related discussion; and finally Sect. 8 summarises the conclusions and discusses future work.

2   Background and Related Work

This section reviews existing research to establish the current state of the art in SH sensor systems, as well as determine the strengths and limitations of the machine learning algorithms and adaptive environments used.

2.1 Smart-Home Sensor Systems

SHs can consist of a variety of different sensors ranging from ambient sensors, body-sensor networks, to video-based solutions [6]. Ambient wireless sensor systems consist of sensors embedded in the environment of the user such as Passive Infrared (PIR) motion, magnetic contact, temperature, and LUX sensors [7]. Generally, a large number of ambient sensors are required to be present in the room in order to track an activity [8], however they are viewed as less intrusive and more acceptable than other types of sensing techniques which include video monitoring [9]. Body-sensor systems include devices that are physically worn or carried by the user such as fitness monitors and smartphones [10]. These can provide physiological data such as heart-rate and body temperature, along with activity data through embedded accelerometer, gyroscope, etc. The downside of body sensors is that they are often viewed as intrusive and older adults may forget to wear them every day [6]. Video-based methods generally provide the most contextual information for HAR, but are viewed as less acceptable due to security and privacy issues [11].

2.2   Machine Learning Algorithms for Activity Recognition Systems

Offline Learning HAR systems

In activity recognition, semi-supervised and unsupervised algorithms are generally deployed for discovering activities from unlabelled data in an offline system. Kwon et al. [12] present a comparison of unsupervised clustering techniques which can be used to discover activity clusters in unlabelled sensor data. Their research focused on recognising activities from smartphone data such as walking, sitting, standing and running. The unsupervised machine learning techniques included Gaussian mixture models, Hierarchical Agglomerative Clustering (HAC), DBSCAN, and K-means clustering. HAC and DBSCAN achieved the highest accuracy and the authors noted HAC for its flexibility and ability to discover hierarchies in the data. HAC works by initially assigning each data point as a cluster, then merging the clusters together at each step. A cut-off can be set as to when the merging should stop, and the various steps of the clustering process can be viewed in the form of a dendrogram.

Real-time Learning HAR and Adaptive Environments

Yala et al. [13] present two methods that utilize an incremental support vector machine (SVM) along with clustering techniques to perform online HAR from streaming sensor data. They utilize a sliding sensor window that encodes the preceding sensor events in order to classify them. Other researchers have also experimented with similar sliding window techniques, ranging from static length sensor windows to dynamic time windows [14].

Drawbacks of traditional supervised HAR techniques are well summarised in [15], in which Ntalampiras and Roveri propose an interesting framework that utilizes Hidden Markov Models (HMM) for performing HAR and incremental dictionary learning to account for new user activities. The only assumption made in this framework is that the activities last for a specific amount of time. The researchers evaluated their technique on gyroscope and accelerometer data from activities such as walking, sitting and standing with promising results.

Most HAR approaches reviewed have utilized supervised learning or have been focused on simple physical activities. This paper instead proposes a semi-supervised framework for recognising and modelling more complex activities of daily living such as cooking, whilst also adapting to changing user behaviour.

3   Problem Formulation

There are two problems the research presented in this paper aims to solve – performing HAR from unlabelled data and tracking user activity in real-time. Traditional HAR approaches train a classifier on datasets which include a labelled dictionary of activities: $\mathcal{A} = \{A_1,...A_N\}$, where A_i denotes the i-th activity and N is the total number of activities. The classifier is then utilized to classify new occurrences of activities as one that is present in $\mathcal{A}$. This method has shown promising results, but does not account for any new activities that the user may start performing which were not present in the original dictionary or changes over time in the way the user performs those activities.

The second problem involves tracking the user activity which is crucial to providing real-time support. To formulate this problem, we first define the term user activity (or task) as a combination of multiple sets of user actions (or subtasks): $A = \{S_1,...S_N\}$, where S_i denotes the i-th subtask and N is the total number of subtasks. The sequence in which these subtasks are performed may vary, as there may be multiple routes to the end goal of the task [16]. The challenge involved in tracking user activity is modelling the relationships between subtasks present in the activity, in order to predict the next sequence of subtasks.

Therefore, the goal of this research is to present a framework which can - automatically generate a dictionary of frequently repeated activities (defined in the next section) by analysing unlabelled sensor data; modify this dictionary according to changes in user behaviour; and model the relationships between the subtasks for each activity to allow for real-time tracking and support.

4   Adaptive Activity Recognition Framework

This section describes the overall framework of the proposed HAR system which aims to identify and model frequently repeated activities performed by the user in order to set up automated assistance/prompts and continually track changes in the performance of these activities for health assessment purposes. The system utilizes ambient and passive sensors embedded throughout the house rather than wearables or video data due to their greater acceptability as established in the literature review.

We first introduce a short-term system memory for storing and analysing sensor data. The system memory is specified as 'n' days and is the number of preceding days's data the system would use from which to identify activities. If the present day is p, then a repeating activity will be identified as such if it is repeated at least once since day p-n. The value of n can be set to a larger integer for tracking longer-term activity model changes, but a smaller value of n favours a more adaptive system. A real-world implementation would require multiple non-overlapping system memories of different sizes to keep track of weekday and weekend behaviour separately (as these might be quite different). This paper only evaluates HAR of weekday behaviour.

The HAR system is divided into two parts, offline HAR and online HAR. Each of these parts are comprised of three steps (Fig. 1). The purpose of the offline HAR system is to discover activity clusters present in unlabelled user data at the end of each day and compare it to the activity clusters found in the previous n days to identify any repeating activities. This is then used to build a model of each identified repeating activity which can be used for performing online or real-time HAR for automated assistance/prompts and updating the activity model based on user actions.

These steps are repeated every day for the database of activity models to be created and updated. Any data older than the specified system memory is deleted, which results in old activities/habits being removed from the database. Any newly discovered repeating activities are added, while existing activity models are updated based on user behaviour (Fig. 2).

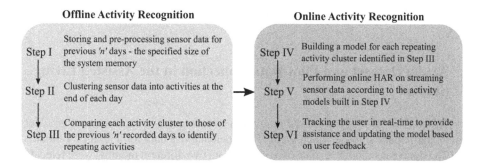

Fig. 1. Adaptive human activity recognition framework

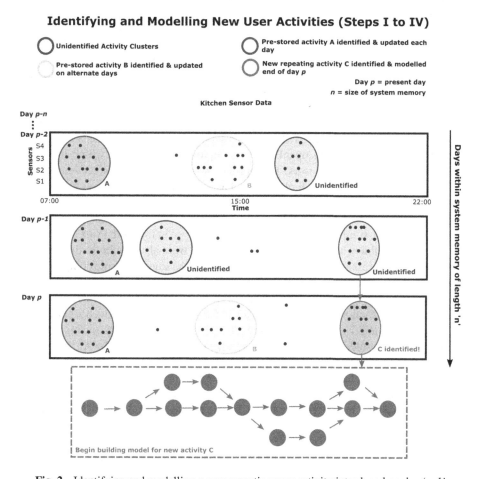

Fig. 2. Identifying and modelling a new repeating user activity introduced on day '*p-1*'

The following section describes the datasets used to develop and evaluate the system while Sect. 6 describes each step in more detail along with an explanation of the feature selection and the artificial intelligence implemented for each of the steps in this framework.

5   Aruba Dataset and Pilot Data Collection in the Assisted Living Studio

The Aruba CASAS study deployed multiple motion sensors in each room of an older female participant's house which yielded labelled activity data for a total of 7 months [5, 17]. Four weeks of kitchen data (two weeks from two different months) were selected at random from this dataset, split into two weeks each for training/developing and testing the offline HAR presented in this paper. Kitchen data was selected so that it could be augmented by additional complementary data collected in the Assisted Living Studio (ALSt) kitchen in the Bristol Robotics Laboratory. Weekend data was removed as only the weekday data was used in this study as explained in *3.1*. While the presence of real-world noise in the Aruba dataset provided the ecological validity for testing the offline HAR component of this study, the sensor configuration lacked inclusion of other sensors such as contact sensors on drawers/cupboards, rendering the data insufficient to test the online HAR system presented in this paper. As such, a sample dataset was collected in the ALSt in the Bristol Robotics Laboratory (BRL) from participants recruited to perform typical kitchen-based activities, which served to augment the Aruba dataset. A Z-wave sensor network using openHAB was deployed in the ALSt which included Fibaro FGMS-001 motion sensors, Everspring SM810 magnetic contact sensors on cupboards, and TKB TZ69E wall plugs (to act as a sensor for the kettle). A total of six activity sessions were recorded from four participants on different days. Due to the nature of sensors used in the ALSt data collection, the sensor values only consisted of ON or OFF (0 or 1) values similar to the motion sensors deployed in the Aruba data collection. The number of ambient sensors deployed in the ALSt were also similar to the number of motion sensors present in the various rooms in the Aruba study to ensure compatibility between the two datasets.

The non-scripted kitchen activity selected for this study was a combination of two interleaved sub-activities - making a hot drink (such as tea, coffee or hot chocolate) and preparing a sandwich. The participants were provided with various items and ingredients appropriate for the selected activity and were asked to place these in the ALSt kitchen in a manner they were most comfortable with. They had access to two cupboards, one drawer, a counter-top, fridge, along with a kettle for boiling water and relevant utensils.

In the selected Aruba dataset there was a 30 to 60 min gap everyday between 16:00 and 17:00 after removing unlabelled activities, which was replaced by an ALSt participant's activity session data (Fig. 3). This was done to introduce real-world noise and data from the Aruba dataset, with the combined Aruba-ALSt data used for testing the offline HAR system, and the ALSt activity used for testing the online HAR system.

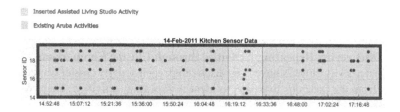

Fig. 3. ALSt activity inserted into Aruba CASAS dataset

6 Pilot Evaluation Study

This section describes an implementation of the adaptive HAR framework along with the data processing and machine learning techniques used. The weekday period was set to 5 days for pilot testing with the combined Aruba-ALSt data. Matlab was used for all data processing.

Step I: Data Pre-Processing
The sensor data obtained from the user for the specified system memory was separated by days and grouped by rooms –

Separation of Days. This was performed so that the sensor data for each day can be processed separately, and then compared to that of the previous day to identify repeating activities.

Grouping Sensors by Rooms. An assumption this paper made was that the activities are restricted by room, which would mean that if a single activity took place over three rooms (such as cleaning), it would be split into three activities (such as cleaning room 1, cleaning room 2, and cleaning room 3). The sensor data was separated and stored accordingly.

Step II: Discovering User Activities - Hierarchical Agglomerative Clustering
The purpose of this step was to discover user activities present in the data at the end of each day. This paper implements HAC to achieve this as used by [12], as it has been proved to be useful for discovering hierarchies of subtasks present within a task. The features used for clustering were the normalised sensor IDs and the timestamp at which the activity occurred. The maximum variance in the data was along the time axis, so HAC would cluster together data points that occurred within short time intervals. The distance calculated between data points was Euclidean. The linkage method was set to 'single', so that the shortest distance was used when clustering. A general rule for the cut off for HAC is to divide the square root of the number of data points by two [18]. After the analysis of the generated dendrograms from the training data, the cut-off formula was changed to

$$Cutoff = \frac{\sqrt{no.\ of\ sensor\ events}}{1.5} \tag{1}$$

Therefore, the cut-off was set dynamically for each day depending on the total number of sensor events present.

Step III: Discovering Repeating Activities through Cluster Comparison
Once distinct activity clusters have been discovered, each activity cluster was compared to clusters from the previous five days to identify repeating activities. There were two features used for this –

1. *The time of the activity cluster.* Calculated as follows –

$$Activity\,time = \frac{time\,of\,last\,sensor\,event - time\,of\,first\,sensor\,event}{2} \tag{2}$$

   This feature was used to compare the time at which the activity occurred on different days. This means that an activity had to be repeated within the same time frame each day for it to be considered a repeating activity (set as one hour in this study after training data analysis). This is a shortcoming of this current implementation of the Adaptive Activity Recognition Framework which could be overcome by a future implementation of dynamic time warping for cluster comparison [19].
2. *The total number of sensor events present in the activity cluster.* This feature was used to represent the sizes of the activity clusters for comparison. In this paper clusters greater than 0.75 and smaller than 1.67 of the original were considered similar (thresholds set after analysis of training data).

Sensor data belonging to activity clusters identified as similar/repeating were stored using a unique identifier to refer to the activity. The data from each identifier was then used to build the activity model for that identifier in the next step.

Step IV: Building User Activity Model using Bayesian Networks
Sensor data from previous days' of user activity were used to generate a Bayesian net of the activity (Fig. 4). This involves identifying and collapsing repeating sequences of sensor events from each previous day. Figure 4 encodes all possible sequences of events the user has performed for the activity in the last five days. Days 3, 4 and 6 are empty as the activity from those days can be reproduced using the sequence of events from days 1, 2 and 5. The probability of the user performing the next sequence of events depends on how many times the user has repeated the same sequence in the past.

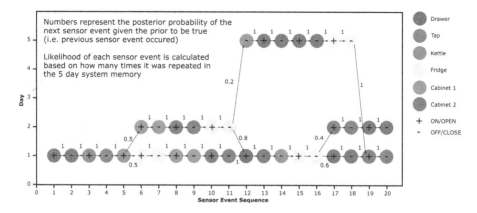

Fig. 4. Bayesian network for assisted living studio activity for participant B

Step V: Real-Time HAR Using Sliding Sensor Window

Once the Bayesian model is generated, a sliding window comprising of three sensor events is used for performing real-time HAR, similar to [13] as shown in Fig. 5 (Note: this is different from the system memory described in Sect. 3.1). This is achieved by first locating the unique activity models from previous days present in the database (generated in step IV) which occurred within an hour of present time. Then comparing the most recent three sensor events (sliding window size = 3) to the first three sensor events present in the activity models to identify the activity being performed.

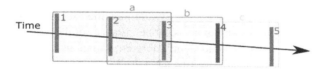

Fig. 5. Sliding window for sensor events – numbered lines are sensor events, lettered boxes are sliding windows of 3 sensor events.

Step VI: Updating User Activity Model through User Feedback

Following real-time recognition, the SH algorithm needs to track the user in to order detect any change in behaviour and automate prompts accordingly. For this, a Matlab script was created to use the Bayesian nets created in the previous step to predict the next most probable move by the user. If the user misses a step, the script keeps track of the user by jumping ahead to reach the step the user is at. However, the script also keeps track of any missed steps in order to provide a prompt to the user to remind him of the missed step. According to the user's response of either correcting for the missed step by doing specific missing action, responding to the prompt with text input, or ignoring the prompt, the Bayesian net for the activity is updated accordingly.

7   Results and Discussion

HAC (Step II) was evaluated on 12 days of Aruba weekday data for kitchen activities between the dates 3rd January 2011 to 18th February 2011 with weekend data removed, as this study only evaluated the weekday system memory. Recognised activities were separated into activities identified precisely and activities identified with noise. An activity was classified as identified precisely if the number of sensor events in the identified activity cluster were same as the number of events listed in the dataset. Some activity clusters contained additional sensor events not part of the label which were classified as "identified with noise" (Fig. 6). The accuracy was calculated as follows

$$Accuracy = \frac{Number\ of\ activities\ identified\ precisely}{Total\ number\ of\ activities\ labelled\ in\ dataset} \times 100 \tag{3}$$

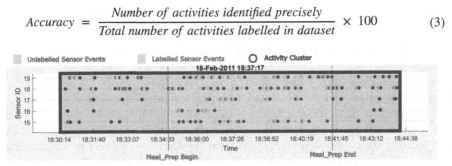

Fig. 6. "Identified with Noise" meal preparation activity, 18 Feb 2011

The overall accuracy of the HAC was 71% for the Aruba dataset in separating sensor events into separate activity clusters. The system identified six separate "Meal Prep" clusters as there were a maximum of six separate instances of the "Meal Prep" activity present in the same day, which could signify different variations of the activity such as preparing different types of meals, making a hot drink, etc.

Cluster comparison (Step III) was then performed to identify the same activities repeating on different days. The system compared the six "Meal Prep" clusters from step II to identify repeating instances of the activity present on other days with an F1-score of 0.93, while the score for the identifying repeating "Wash Dishes" activity was 0.35. For comparison, authors in [13] used the Aruba dataset to achieve an F1-score between 0.6 and 0.7 for "Meal Prep" activities and 0 for "Wash Dishes". They concluded that the classifier failed to identify the "Wash Dishes" activity due to imbalance in the dataset because of the presence of unlabelled sensor events that dominates it. This also seems to be the case here as the "Wash Dishes" activity was performed less consistently than "Meal Prep" in the selected data and the system often classified unlabelled data as "Wash Dishes" during cluster comparison, resulting in the low F1-score. It must also be noted that authors in [13] used six weeks of Aruba dataset including training and testing, while this paper only used 12 days for both, making the former's approach more robustly tested. The system also identified instances of unlabelled data as an activity which were removed from the results as they cannot be verified due to missing labels (Fig. 7).

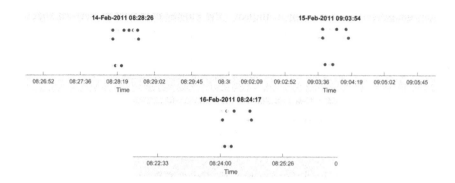

Fig. 7. An unlabelled repeating activity cluster identified and isolated

The F1-score for the cluster comparison of the ALSt activity from the combined Aruba-ALSt dataset was 0.96. Real-time activity recognition (Step V) for the ALSt participant activity from day 2 onwards occurred at the beginning of the activity (3rd sensor event) 64% of the time, and near the midway point (5th to 7th sensor event) 96% of the time. On day 6, real-time activity recognition occurred at the beginning of the activity 100% of the time for all participants. This is because the Bayesian activity model improves each day utilizing the previous 5 days of data. It must also be noted that the earliest real-time detection can only happen at the 3rd sensor event due to the size of the sliding sensor window being three sensor events.

The last experiment was to evaluate how well the system can track the user to detect missed steps using the Bayesian network and provide prompts (Step VI), which was achieved by simulating the user forgetting to open/close the cabinet and the fridge during the activity on day 5 and day 6. The system detected these missed steps in 75% of the instances on day 5 and 87.5% of the instances on day 6.

8 Conclusions and Future Work

A significant problem with current HAR techniques is the requirement of large amounts of expertly labelled training data for each specific end-user. This paper presented an alternative approach to HAR which utilises a short-term system memory with specified decay length to recognise repeating activities from unlabelled data and track them in real-time. The user activity models generated by this system can be utilized to offer additional support, for example, if the user forgets or misses a step in a known activity pattern such as forgetting to turn off the hob while preparing a meal, the system can then remind them through automated prompts. Additionally, this approach can be used to track changes in user activities which could be an indication of deteriorating health conditions and alert the carer of significant changes based on a personalised threshold. An implementation of the proposed HAR framework was also pilot tested using the Aruba dataset augmented with sample data collected in the ALSt in the Bristol Robotics Laboratory with promising results. Future work will include testing this framework on different datasets and attempting to improve the cluster comparison performance by implementing techniques such as dynamic time warping [19].

References

1. Demiris, G., Hensel, B.K.: Technologies for an aging society: a systematic review of "smart home" applications. Yearb Med. Inf. **3**, 33–40 (2008)
2. Kvedar, J., Coye, M.J., Everett, W.: Connected health: a review of technologies and strategies to improve patient care with telemedicine and telehealth. Health Aff. (Millwood) **33**, 194–199 (2014)
3. Shoaib, M., Bosch, S., Incel, O., Scholten, H., Havinga, P.: A survey of online activity recognition using mobile phones. Sensors **15**, 2059–2085 (2015)
4. Ranasinghe, S., Al MacHot, F., Mayr, H.C.: A review on applications of activity recognition systems with regard to performance and evaluation. Int. J. Distrib. Sens. Netw. **12**, 1550147716665520 (2016)
5. CASAS: Smart Home Projects, Washington State University (USA). http://casas.wsu.edu/
6. Pal, S., Feng, T., Abhayaratne, C.: Real-time recognition of activity levels for ambient assisted living. In: 5th IEEE International Conference Consumer Electronics - Berlin ICCE-Berlin 2015, pp. 485–488 (2016)
7. Roggen, D., Förster, K., Calatroni, A., Tröster, G.: The adARC pattern analysis architecture for adaptive human activity recognition systems. J. Ambient Intell. Humaniz. Comput. **4**, 169–186 (2013)
8. Cardinaux, F., Bhowmik, D., Abhayaratne, C., Hawley, M.S.: Video based technology for ambient assisted living: A review of the literature. J. Ambient Intell. Smart Environ. **3**, 253–269 (2011)
9. Basu, D., Moretti, G., Sen Gupta, G., Marsland, S.: Wireless sensor network based smart home: Sensor selection, deployment and monitoring. In: 2013 IEEE Sensors Applications Symposium, SAS 2013 – Proceedings, pp. 49–54 (2013)
10. Rashidi, P., Mihailidis, A.: A survey on ambient-assisted living tools for older adults. IEEE J. Biomed. Heal. Inform. **17**, 579–590 (2013)
11. Boise, L., Wild, K., Mattek, N., Ruhl, M., Dodge, H.H., Kaye, J.: Willingness of older adults to share data and privacy concerns after exposure to unobtrusive home monitoring. Gerontechnology **11**, 428–435 (2013)
12. Kwon, Y., Kang, K., Bae, C.: Unsupervised learning for human activity recognition using smartphone sensors. Expert Syst. Appl. **41**, 6067–6074 (2014)
13. Yala, N., Fergani, B., Fleury, A.: Feature extraction and incremental learning to improve activity recognition on streaming data. In: 2015 IEEE International Conference on Evolving and Adaptive Intelligent Systems EAIS 2015, pp. 1–8 (2015)
14. Krishnan, N.C., Cook, D.J.: Activity recognition on streaming sensor data. Pervasive Mob. Comput. **10**, 138–154 (2014)
15. Ntalampiras, S., Roveri, M.: An incremental learning mechanism for human activity recognition. In: 2016 IEEE Symposium Series on Computational Intelligence, pp. 1–6 (2016)
16. Storf, H., Kleinberger, T., Becker, M., Schmitt, M., Bomarius, F., Prueckner, S.: An event-driven approach to activity recognition in ambient assisted living. In: Tscheligi, M., et al. (eds.) AmI 2009. LNCS, vol. 5859, pp. 123–132. Springer, Heidelberg (2009). https://doi.org/10.1007/978-3-642-05408-2_16
17. Cook, D.J.: Learning setting- generalized activity models for smart spaces. IEEE Intell. Syst. **27**, 32–38 (2012)

18. Duda, R.O., Hart, P.E., Stork, D.G.: Pattern Classification, 2nd edn. John Wiley & Sons, Hoboken (2001)
19. Petitjean, F., Ketterlin, A., Gançarski, P.: A global averaging method for dynamic time warping, with applications to clustering. Pattern Recogn. **44**, 678–693 (2011)

Structured Inference Networks Using High-Dimensional Sensors for Surveillance Purposes

Vincent Polfliet[1], Nicolas Knudde[1]($\boxtimes$), Baptist Vandersmissen[2],
Ivo Couckuyt[1], and Tom Dhaene[1]

[1] Department of Information Technology, Ghent University - imec, Ghent, Belgium
nicolas.knudde@ugent.be
[2] Department of Electronics and Information Systems,
Ghent University - imec, Ghent, Belgium

Abstract. Video cameras are arguably the world's most used sensors for surveillance systems. They give a highly detailed representation of a situation that is easily interpreted by both humans and computers. However, these representations can lose part of their representational value when being recorded in less than ideal circumstances. Bad weather conditions, low-light illumination or concealing objects can make the representation more opaque. A radar sensor is a potential solution for these situations, since it is unaffected by the light intensity and can sense through most concealing objects. In this paper, we investigate the performance of a structured inference network on data of a low-power radar device. A structured inference network applies automated feature extraction by creating a latent space out of which the observations can be reconstructed. A classification model can then be trained on this latent space. This methodology allows us to perform experiments for both person identification and action recognition, resulting in competitive error rates ranging from 0% to 6.5% for actions recognition and 10% to 12% for person identification. Furthermore, the possibility of a radar sensor being used as a complement to a camera sensor is investigated.

Keywords: Structured inference network · Person identification
Action recognition · Indoor sensing · Micro-doppler · Sensor fusion

1 Introduction

In recent years, interest in autonomous surveillance systems grew considerably. While these systems improved significantly, the primary sensors remained the same. The dominance of video cameras in autonomous surveillance systems can be explained by their fundamental strengths. They give a detailed high dimensional representation of their environment, which is easily interpretable by humans as well as computers. Moreover, reduction in price and higher resolutions kept driving their success. While the fundamental advantages of video

© Springer Nature Switzerland AG 2018
E. Pimenidis and C. Jayne (Eds.): EANN 2018, CCIS 893, pp. 16–27, 2018.
https://doi.org/10.1007/978-3-319-98204-5_2

cameras were a big catalyst for the early development of surveillance systems, their deficiencies are now holding them back. For some of these deficiencies, such as recordings in bad weather conditions or low light environments, workarounds can be found. Others such as concealing clothing are harder to deal with. In contrast, a radar sensor is unaffected by concealing clothing, bad weather conditions, low-light environments and can be placed out of sight, behind a wall.

A radar is an active sensor that transmits an electromagnetic signal, which is reflected by objects in its line of sight. Information about these objects is then extracted out of these signals taking advantage of, e.g. the Doppler effect. Moreover, individual moving parts of a person or object will each reflect their own Doppler signal which are then summarized into a micro-Doppler (MD) signature [3].

These signatures contain information about the movement of the target, providing a promising feature to differentiate between for example cars, bikers, pedestrians or dogs. Another use for these MD signatures is to recognise different actions, ranging from walking to sitting or boxing [10,11]. However, perhaps the most challenging application is to differentiate individuals based on the way they move, the so called gait-based identification. While there is a noticeable difference between how a dog and a human walks or how a person runs or sits, the difference in the MD signature between two persons walking is more subtle. This subject has been extensively researched, however, previous papers used a high-power radar sensor with relatively simple scenarios. In this paper the data sets are recorded using a low-power frequency modulated continuous wave (FMCW) radar. This radar is a low-cost, power efficient and compact sensor suited for indoor usage. However, the combination of a human's low radar cross-section and a low-power device poses a significant challenge for this study [4].

Two data sets are used for our experiments. The first uses the IDentification with Radar (IDRad) benchmark, which is an extensive data set where the main objective is to identify individuals moving randomly in a room [19]. An additional data set is recorded where the main objective is to recognise different actions. Previous studies applied either deep convolutional neural networks (DCNN) [11,19] or clustering methods [9,20] to MD signatures. Both approaches were successful by exploiting certain properties of the data. The DCNN tries to take advantage of the spatial properties, along the time and velocity axes, of an MD signature. Conversely, the clustering methods are applied on feature vectors of the original noisy data. Hence, a structured inference network (SIN) [14] can potentially exploit both these properties due to its inherent Markovian properties. This model creates a lower dimensional latent space into which each time step is projected without losing their sequential dependencies. The lower dimensional states also implies that the model performs autonomous feature selection on the data. The resulting lower dimensional latent states are then used in a classification model. These properties make the SIN well-suited for high dimensional sequential data, such as radar data.

2   Related Work

There has been extensive research in the use of radar as a sensor. This section will highlight several relevant studies concerning action recognition and person identification. Afterwards some other recent results will be discussed regarding SINs.

Action recognition and *gait-based identification* are discussed in a wide array of studies. The former is usually defined by the amount of different actions in the data set. In [10,11], 7 actions are proposed, ranging from walking, walking with a stick, running to even boxing. A wide variety of models have been investigated to differentiate between actions. Kim et al. apply a support vector machine with manual engineered features [10] and an DCNN [11]. In [16], transfer learning is applied to a pretrained CNN. In [5], singular value decomposition with multiple classification models were used for detecting violent intents. The studies [7,15], investigate autonomous surveillance systems as a tool to monitor elderlies using a wide array of classifiers.

Conversely, mainly data driven models are studied for gait-based identification. In [6] k-means and k-NN clustering is used on thirteen subjects with an accuracy ranging from 92.4% to 100%. The authors of [17] also apply k-NN along with two manual engineered features and Kalgaonkar and Raj obtained an accuracy of 90% by using a Gaussian mixture model (GMM) [9]. Finally, the authors of [19] designed a deep convolutional neural network (DCNN) resulting in an accuracy of 81.61% on lower-power radar data.

Radar data can also be used for non-classification purposes such as person tracking [13].

The structured inference network was proposed in [14]. The authors apply the model to the reconstruction of polyphonic music and the counterfactual prediction of electronic health records of patient data. This model was then also used by the authors of [18] to model human poses. A similar black box variational inference model for state space models is proposed in [2]. While an unsupervised model is proposed in [8], which combines the strengths of a latent graphical variational auto-encoder (VAE) and GMM by using a conditional random field as their inference network. The authors apply their model to a data set of a mouse running in a box, where it successfully clusters different movements of the mouse.

3   Micro-doppler

A large object or body moving through a room at a constant speed induces a constant Doppler Frequency shift. However, smaller moving parts can cause additional micro-motion dynamics, which, in their turn, induce Doppler modulations on the echoed signal. This is referred to as the micro-Doppler effect [3] and causes sidebands around the Doppler frequency, representing the different smaller moving parts. The micro-Doppler map can thus be seen as the power reflected as a function of the speed of the reflector. The radar used in the data sets is a 77 GHz Frequency Modulated Continuous Wave radar. An FMCW radar has the advantage of being power efficient, but comes as the expense of a low signal to noise ratio, which makes analysing this sensor data more challenging.

4   Structured Inference Network

A structured inference network [14] is a subfield of machine learning where it is assumed that the data confirms to the structure of a *Gaussian state space model* (GSSM). A GSSM assumes that the actual states of a situation are only partly observable and that there exist latent states that fully describe the context of the data without any error. These states are then also assumed to be continuous and only dependent on their previous state. However, in data-oriented problems, the parametric form for a GSSM is usually unknown. A solution for this is a *deep Markov model* (DMM): A GSSM where the emission and transition functions are replaced by multi-layer perceptrons (MLP). The resulting GSSM still has the Markovian structure of an hidden Markov model (HMM) but uses the strength of deep neural networks to help model complex data. An example of a DMM can be seen in Fig. 1.

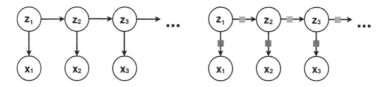

Fig. 1. Generative models of sequential data: **(left)** is a classical HMM. While **(right)** depicts a DMM. The transition (green) and emission (red) functions are both approximated using MLPs. (Color figure online)

The model requires that the latent states are multivariate Gaussian distributions, with a mean and covariance that are functions dependent on the previous latent state. In this paper we also define our observations to be multivariate Gaussian distributions where the parameters are dependent on the latent state. Equation 1 results in a GSSM with model parameters $\theta = \{\alpha, \beta, \kappa, \lambda\}$.

$$
\begin{aligned}
\mathbf{z}_t &\sim \mathcal{N}(G_\alpha(\mathbf{z}_{t-1}, \Delta_t), S_\beta(\mathbf{z}_{t-1}, \Delta_t)) \quad (Transition) \\
\mathbf{x}_t &\sim \mathcal{N}(G_\kappa(\mathbf{z}_t, \Delta_t), S_\lambda(\mathbf{z}_t, \Delta_t)) \quad\quad (Emission)
\end{aligned}
\tag{1}
$$

Another technique needed for this model is *variational learning* [12]. Assume that $p(\mathbf{x}, \mathbf{z}) = p_\theta(\mathbf{z})p_\theta(\mathbf{x}|\mathbf{z})$ is a generative model, where $\mathbf{x}$ is the observation and $\mathbf{z}$ the latent variable. The posterior distribution for this generative model is usually intractable. The variational principle then states that there should be an approximation of the posterior distribution $q_\phi(\mathbf{z}|\mathbf{x})$. Using this principle, a lower bound of the marginal likelihood is found, which is parameterized by a neural network.

$$\log p_\theta(\mathbf{x}) \geq \mathop{\mathbb{E}}_{q_\phi(\mathbf{z}|\mathbf{x})} [\log p_\theta(\mathbf{x}|\mathbf{z})] - \mathrm{KL}(q_\phi(\mathbf{z}|\mathbf{x})||p_\theta(\mathbf{z})) \tag{2}$$

Using variational learning a lower bound is found that approximates the posterior distribution of the GSSM [14].

$$\mathcal{L}(\mathbf{X}; (\boldsymbol{\theta}, \boldsymbol{\phi})) = \sum_{t=1}^{T} \mathop{\mathbb{E}}_{q_\phi(\mathbf{z}_t|\mathbf{X})} [\log p_\theta(\mathbf{x}|\mathbf{z})] - \mathrm{KL}(q_\phi(\mathbf{z}_1|\mathbf{X}||p_0(\mathbf{z}_1))$$
$$- \sum_{t=2}^{T} \mathop{\mathbb{E}}_{q_\phi(\mathbf{z}_{t-1}|\mathbf{X})} [\mathrm{KL}(q_\phi(\mathbf{z}_t|\mathbf{z}_{t-1}, \mathbf{X})||p_\theta(\mathbf{z}_t|\mathbf{z}_{t-1}))] \tag{3}$$

Since the latent states of the generative model will be used for classification purposes, we propose an additional modification. By using a different prior for each classification target, we can encourage the latent state to be more accommodating regarding the classification.

5   Methodology

The main objective of this paper is to investigate the efficiency of a SIN applied to MD signatures for two use cases: gait-based person identification and action recognition. Both data sets were recorded using the same low-power FMCW radar, produced by INRAS [1], in an empty indoor environment. The action recognition data set was recorded to study the performance of radar sensors versus camera sensors.

5.1   Preprocessing

Radar: The MD signature is first achieved by calculating a two-dimensional Fourier transform on the range-Doppler map. Afterwards the absolute values are converted to decibels and are summed over the range dimensions. The raw MD signature contains 256 Doppler channels per time step (with 15 fps). Each of these channels represents a speed ranging from -3.8 m/s to 3.8 m/s. The static channels representing the highest and lowest speeds are removed, without any loss of relevant information. Subsequently, the resulting sequence is thresholded by fixing every point under a certain value. After thresholding, a logarithmic scaling step is applied to compress high activated values, which results in a lower variance. Finally, each Doppler channel will be normalized separately for each sequence. Figure 2 displays the different results of the preprocessing steps to transform a raw MD signature to the fully preprocessed MD signature.

Camera: As the video camera data is only used for basic action recognition, there is no need for highly detailed images. Taking this in consideration, the images were first converted to gray scale and then rescaled from 640×480 pixels to 30×20 pixels. The resulting images are then normalized using the mean pixel values. Finally, the camera images are processed by a small convolutional network, as shown in Fig. 4. A partial copy of the camera data set was also created with half of the image occluded (left side). This area will serve as an artificial screen to check the performance between a camera sensor and a radar sensor in less than ideal circumstances. The intermediate results of the preprocessing and an example of an occluded image are shown in Fig. 3.

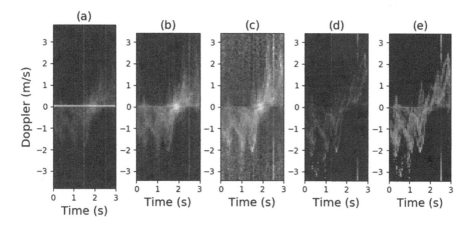

Fig. 2. A 3 s MD signature, each figure shows the results of a preprocessing step, with first **(a)** showing the raw signature. **(b)** is then obtained by removing the static channels. **(c)** is the normalized MD signature of **(b)** and still displays a lot of noise. This is then solved by applying thresholding **(d)** and finally the variance in the high activated areas is reduced by log scaling **(e)**.

Fig. 3. Camera images from the Actions data set: From left to right we have the raw image **(a)**, conversion to gray scale with rescaling **(b)**, normalized image **(c)** and the occluded version of the image **(d)**.

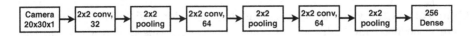

Fig. 4. Convolutional neural network to compress the camera images to lower dimensional vectors.

Sensor Fusion: The high-dimensional radar and camera data are represented by vectors after their respective preprocessing steps. A straightforward form of sensor fusion is obtained by concatenating them. However, both vectors might contain duplicate information. This is filtered out by sending the concatenated vectors through a dense layer. The resulting vector can then be mapped by the SIN to obtain a latent space containing the information of both sensors.

5.2   Model

We implemented the SIN, using the theory mentioned in Sect. 4, in Tensorflow. An outline of the model is shown in Fig. 5. The data will be fed into the Recurrent Neural Network (RNN), which is used as a generative model to create the latent space. Afterwards these states will go through the emission and transition MLPs to find a prediction for respectively the observations and the next latent state. These three predictions and the actual data are then used to calculate the likelihood. Once the SIN is trained, a classification model is applied on the latent states from the generative model. Three different classification models were tested and can be seen in Fig. 6.

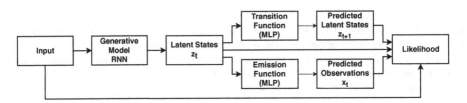

Fig. 5. An outline of the SIN: A generative model is coupled with two MLPs that represent the transition and emission functions.

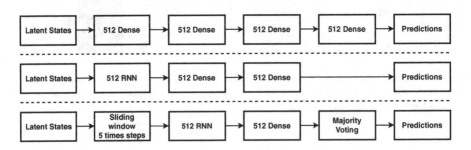

Fig. 6. Three different possible classification models from top to bottom: a MLP, a RNN, and a RNN with majority voting.

6   Experiments

First the efficiency of the model for gait based identification is investigated using the IDRad data set. Afterwards the results of the action recognition data set will be discussed, comparing both camera and radar sensors.

6.1   Person Identification

The IDRad contains recordings of 5 people. Each test person was required to walk for 20 min in random directions with abrupt stops and turns in 2 empty rooms. Each model is trained using sequences of 3 s, which allows us to compare our results with [19].

Analysis of the Generative Model: The classification models are trained on the latent space created by the SIN. However, this model is trained on the likelihood of the reconstruction of the data and is thus independent of the targets of the data. This means that the performance of the classification depends on how well the SIN generalizes the latent space regarding the classification, making the training time of the SIN a hyperparameter. Figure 7 shows the impact of the training time of the SIN on both the classification loss as well as the reconstruction likelihood. While the structured inference network keeps improving over time, the classification model reaches its peak performance in the 100 to 200 epochs interval.

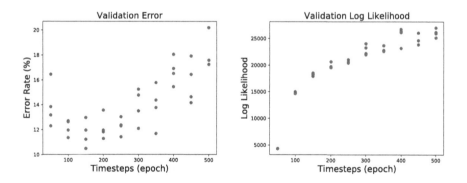

Fig. 7. These figures display the impact of the training time of the SIN on the validation error rate of the classification and on the validation log-likelihood of the SIN itself. It can be seen that while the log-likelihood keeps improving over time, this is not the case for the classification error. The best performing classification models are thus trained on the latent states of SINs with a training time between 100 to 200 epochs.

Results: The structured inference network was trained for 150 epochs and repeated between 10 and 20 times for each experiment.

Table 1 illustrates the impact of the preprocessing. We can see that the results improve when removing the static channels. The results do not improve when adding either the thresholding or log scaling preprocessing step. However when combining both these preprocessing steps, we are able to obtain the best performing models. This is due to the variance of the reconstruction being lowered in the low activated areas by thresholding and in the high activated areas by log scaling. The results thus only improve when both variances are lowered.

Table 2 shows the results of the different classification models on the latent state. Each classification model was tested on the same latent space created by a SIN. Here can be seen that the RNN model outperforms both other models.

Table 1. The impact of adding or removing a subsequent preprocessing step on the error rate. The error rate displays the mean and standard deviation of results over the 5 runs.

Preprocessing	Validation	Test
Raw	$40.90 \pm 3.24\%$	$32.72 \pm 2.21\%$
Remove Static	$21.24 \pm 1.25\%$	$29.59 \pm 1.57\%$
Remove Static + Thresholded	$22.11 \pm 3.13\%$	$32.86 \pm 3.53\%$
Remove Static + Log Scaled	$32.37 \pm 6.54\%$	$38.78 \pm 3.26\%$
Full Preprocessing	$11.92 \pm 0.92\%$	$10.44 \pm 0.76\%$

Table 2. The performance of the different classification models by their error rate.

Classification Model	Validation	Test
MLP	23.14%	18.39%
RNN	11.64%	10.34%
RNN MV	11.78%	10.17%

Table 3. The performance of the two types of structured inference networks and the results of the DCNN as stated in [19]. The error rate displays the mean and standard deviation of results over the 5 runs.

Model	Validation	Test
PCA + RF [19]	48.86%	38.59%
DCNN [19]	24.70%	21.54%
RNN	$15.26 \pm 1.62\%$	$12.20 \pm 0.26\%$
SIN + RNN	$12.24 \pm 1.49\%$	$10.66 \pm 0.74\%$
SIN multiple priors + RNN	$11.92 \pm 0.92\%$	$10.44 \pm 0.76\%$

Finally, Table 3 compares the results found in [19] using a DCNN and principle component analysis with an SVM versus a basic RNN, a SIN and a SIN with different priors. It can be seen that the previous benchmark is improved by up to 12% on the validation set and 11% on the test set using the extra log scaling preprocessing step and a SIN.

6.2   Action Recognition

The data used in this experiment contains radar and camera data of actions generated by 3 people. It consists of 540 samples of 3 s each. Each sample represents either a person walking, sitting down or falling. For these experiments the same optimal preprocessing was used as described in Sect. 5.1.

Correlation Between Camera and Radar Sequences: The structured inference network can be used to check for correlation between the two sensors.

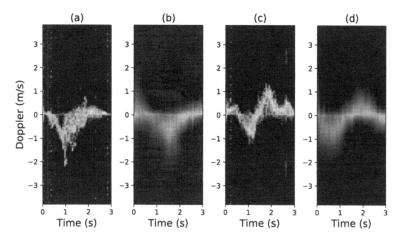

Fig. 8. Artificially generated MD signatures created from camera sequences (b) and (d) versus the original MD signatures (a) and (c).

Table 4. The performance with different sensors by their error rate. SL implies that a screen was artificially inserted on the left side of the camera images, occluding half of the image.

Input	Validation	Test
Radar	3.78 ± 1.29%	6.33 ± 3.13%
Camera	0.11 ± 0.33%	0.67 ± 0.74%
Camera SL	4.78 ± 1.61%	5.56 ± 2.19%
Radar and Camera	0.11 ± 0.33%	0.72 ± 0.73%
Radar and Camera SL	3.33 ± 1.17%	3.39 ± 1.43%

This is done by training the model on the reconstruction of the first sensor's data and using the second sensor as input. The results of reconstructing MD signatures out of camera sequences can be seen in Fig. 8. While these are not exact reconstructions, the shape of the MD signatures are very similar, confirming the correlation that the log-likelihood of the model suggested.

Results: Table 4 shows difference in results between the radar and camera sensor. The camera data performs better than the radar, with an error rate of 0.67% compared to 6.33%. However, the radar data performs equally well when half of the camera image is occluded. Then by combining the radar and camera data, the problem of the screen was partially alleviated resulting in error rates or 3.33%, which is 2 to 3% lower than the individual sensors.

7 Conclusion and Future Work

We propose to use a classification model on top of the latent space created by a structured inference network and show it outperforms previous methods such as a deep convolutional neural network. This is illustrated on novel use cases of high dimensional camera and radar sequences, where we also show its potential to be used for sensor fusion.

It is noted that the performance of the classification model naturally depends on the amount of trained epochs of the structured inference network, since the latent space is created without consideration of the targets. A possible solution for this could be the unsupervised model mentioned in [8], which combines the strengths of a structured variational auto-encoder with a GMM. Another research point is to apply this model on more challenging radar data, such as walking around with an object or walking in a furnished room.

References

1. Inras gmbh (2017). http://www.inras.at
2. Archer, E., Memming Park, I., Buesing, L., Cunningham, J., Paninski, L.: Black box variational inference for state space models. ArXiv e-prints, November 2015
3. Chen, V.C., Li, F., Ho, S.S., Wechsler, H.: Micro-doppler effect in radar: phenomenon, model, and simulation study. IEEE Trans. Aerosp. Electr. Syst. **42**(1), 2–21 (2006). https://doi.org/10.1109/TAES.2006.1603402
4. Chen, V., Tahmoush, D., Miceli, W.: Radar micro-doppler signatures: processing and applications (2014)
5. Fioranelli, F., Ritchie, M., Griffiths, H.: Classification of unarmed/armed personnel using the netrad multistatic radar for micro-doppler and singular value decomposition features. IEEE Geosci. Remote Sens. Lett. **12**(9), 1933–1937 (2015). https://doi.org/10.1109/LGRS.2015.2439393
6. Garreau, G., et al.: Gait-based person and gender recognition using micro-doppler signatures. In: 2011 IEEE Biomedical Circuits and Systems Conference (BioCAS), pp. 444–447, November 2011. https://doi.org/10.1109/BioCAS.2011.6107823

7. Gurbuz, S.Z., Clemente, C., Balleri, A., Soraghan, J.J.: Micro-doppler-based in-home aided and unaided walking recognition with multiple radar and sonar systems. IET Radar Sonar Navig. **11**(1), 107–115 (2017). https://doi.org/10.1049/iet-rsn.2016.0055

8. Johnson, M., Duvenaud, D.K., Wiltschko, A., Adams, R.P., Datta, S.R.: Composing graphical models with neural networks for structured representations and fast inference. In: Lee, D.D., Sugiyama, M., Luxburg, U.V., Guyon, I., Garnett, R. (eds.) Advances in Neural Information Processing Systems 29, pp. 2946–2954. Curran Associates, Inc. (2016). http://papers.nips.cc/paper/6379-composing-graphical-models-with-neural-networks-for-structured-representations-and-fast-inference.pdf

9. Kalgaonkar, K., Raj, B.: Acoustic doppler sonar for gait recoginaton. In: 2007 IEEE Conference on Advanced Video and Signal Based Surveillance, pp. 27–32, September 2007. https://doi.org/10.1109/AVSS.2007.4425281

10. Kim, Y., Ling, H.: Human activity classification based on micro-doppler signatures using a support vector machine. IEEE Trans. Geosci. Remote Sens. **47**(5), 1328–1337 (2009). https://doi.org/10.1109/TGRS.2009.2012849

11. Kim, Y., Moon, T.: Human detection and activity classification based on micro-doppler signatures using deep convolutional neural networks. IEEE Geosci. Remote Sens. Lett. **13**(1), 8–12 (2016). https://doi.org/10.1109/LGRS.2015.2491329

12. Kingma, D.P., Welling, M.: Auto-encoding variational bayes. arXiv preprint arXiv:1312.6114 (2013)

13. Knudde, N., et al.: Indoor tracking of multiple persons with a 77 GHz MIMO FMCW radar. In: 2017 European Radar Conference (EURAD), pp. 61–64, October 2017. https://doi.org/10.23919/EURAD.2017.8249147

14. Krishnan, R., Shalit, U., Sontag, D.: Structured inference networks for nonlinear state space models (2017). https://aaai.org/ocs/index.php/AAAI/AAAI17/paper/view/14215

15. Liu, L., Popescu, M., Skubic, M., Rantz, M., Yardibi, T., Cuddihy, P.: Automatic fall detection based on doppler radar motion signature. In: 2011 5th International Conference on Pervasive Computing Technologies for Healthcare (PervasiveHealth) and Workshops, pp. 222–225, May 2011. https://doi.org/10.4108/icst.pervasivehealth.2011.245993

16. Park, J., Javier, R.J., Moon, T., Kim, Y.: Micro-doppler based classification of human aquatic activities via transfer learning of convolutional neural networks. Sensors. **16**(12), 1990 (2016)

17. Tahmoush, D., Silvious, J.: Radar micro-doppler for long range front-view gait recognition. In: 2009 IEEE 3rd International Conference on Biometrics: Theory, Applications, and Systems, pp. 1–6, September 2009. https://doi.org/10.1109/BTAS.2009.5339049

18. Toyer, S., Cherian, A., Han, T., Gould, S.: Human pose forecasting via deep markov models. arXiv preprint arXiv:1707.09240 (2017)

19. Vandersmissen, B., et al.: Indoor person identification using a low-power FMCW radar. IEEE Trans. Geosci. Remote Sens. **PP**, 1–12 (2018). https://doi.org/10.1109/TGRS.2018.2816812

20. Zhang, Z., Andreou, A.G.: Human identification experiments using acoustic micro-doppler signatures. In: 2008 Argentine School of Micro-Nanoelectronics, Technology and Applications, pp. 81–86, September 2008

Deep Learning

Deep Imitation Learning with Memory for Robocup Soccer Simulation

Ahmed Hussein[1(✉)], Eyad Elyan[1], and Chrisina Jayne[2]

[1] School of Computing, Robert Gordon University,
Garthdee Road, Aberdeen AB10 7QB, UK
a.s.h.a.hussein@rgu.ac.uk
[2] School of Engineering, Computing and Mathematics,
Oxford Brookes University, Oxford OX3 0BP, UK

Abstract. Imitation learning is a field that is rapidly gaining attention due to its relevance to many autonomous agent applications. Providing demonstrations of effective behaviour to teach the agent is useful in real world challenges such as sparse rewards and dynamic environments. However, most imitation learning approaches don't retain a memory of previous actions and treat the demonstrations as independent and identically distributed samples. This neglects the temporal dependency between low-level actions that are performed in sequence to achieve the desired behaviour. This paper proposes an imitation learning method to learn sequences of actions by utilizing memory in deep neural networks. Long short-term memory networks are utilized to capture the temporal dependencies in a teacher's demonstrations. This way, past states and actions provide context for performing following actions. The network is trained using raw low-level features and directly maps the input to low-level parametrized actions in real-time. This minimizes the need for task specific knowledge to be manually employed in the learning process compared to related approaches. The proposed methods are evaluated on a benchmark soccer simulator and compared to supervised learning and data-aggregation approaches. The results show that utilizing memory while learning significantly improves the performance and generalization of the agent and can provide a stationary policy than can produce robust predictions at any point in the sequence.

1 Introduction

Recent years have seen a rise in demand for autonomous intelligent agents. Imitation learning [7] is a promising approach for teaching agents intelligent behaviour by providing demonstrations performed by an expert. Providing demonstrations by performing a task is substantially easier than articulating how the task should be performed and explicitly programming the agents. Moreover, learning from demonstrations is suitable for realistic training scenarios which impose restrictions on learning from experience such as sparse rewards and dynamic environments. Imitation learning approaches commonly deal with demonstrations as

© Springer Nature Switzerland AG 2018
E. Pimenidis and C. Jayne (Eds.): EANN 2018, CCIS 893, pp. 31–43, 2018.
https://doi.org/10.1007/978-3-319-98204-5_3

discrete instances of state and action pairs. Although most autonomous applications involve performing sequences of actions to achieve a goal, most learning methods process instances separately as independent and identically distributed (i.i.d.) samples. These methods rely on the hypothesis that the observed state contains enough information to make an accurate decision; and that performing a series of accurate independent decisions will accumulate to effective behaviour. This hypothesis overlooks the dependencies between actions which can be key in planning long trajectories of actions. This is especially sensitive in imitation learning as the teacher might inherently be relying on memory, without this information being presented to the learning agent. Even if an accurate decision can be made from the current state alone, the teacher might choose a different course of action based on previous experience. If this additional information is not presented to the agent, it won't be able to learn from the demonstrated behaviour [6]. Moreover, even if sampling the training data is dependent on previous actions such as data aggregation methods, the learning algorithm doesn't take temporal relationships between these observations into account. Using memory of past events as context, allows the policy to learn different reactions to similar observations in different point along the trajectory [11]. It is therefore necessary to represent training demonstrations as sequences and learn to reproduce dependent action trajectories.

Recurrent neural networks have shown great success in learning from sequences [2,12]. They capture temporal dependencies by having looping connections so the nodes consider previously processed samples along with new input to produce a decision. However, most RNN applications involve processing the sequence in its entirety before producing a decision or generating an output sequence [19]; which is not suitable for real time autonomous agents. Some applications such as handwritten text generation utilize RNNs to generate a sequence one step at a time [3]. However, these sequences are generated in isolation from other factors while autonomous agents are required to react mid trajectory to dynamic environments. For that, imitation learning requires new RNN based methods that can learn from long sequences of dependent actions and react based on real time observations of the environment.

This paper proposes a novel approach which includes representing demonstrations as sequences of dependent state-action pairs and using a long-short-term-memory network (LSTM) to learn a policy. The LSTM network learns a mapping between states and actions while taking into consideration memory of previous events and actions and the temporal dependencies between these instances. This approach is demonstrated on the "robocup soccer simulator" [9]; a multi-agent soccer simulator. The multi-agent setting provides a dynamic environment for which generating static sequences is not suitable as the policy is required to react to the other agents' actions. This makes the simulator a popular benchmark for intelligent agents. Unlike most machine learning methods, the proposed LSTM network learns from raw low-level sensory data, without the need for engineered feature extraction. Similarly, the policy performs low-level parametrized actions that making a decision as well as predicting continuous values for the actuators simultaneously. Performing sequences of these low level actions makes up the desired behaviour without manually engineering high level strategies. To evaluate the proposed LSTM approach its performance is compared to the hand-crafted teacher policy, and policies learned via neural networks without memory (MLP). To evaluate the generalization ability of RNNs in imitation learning, the proposed approach is further compared to a data aggregation method [16] conducted on the MLP agents.

The remainder of the paper is organised as follows: Sect. 2 reviews related work in the literature. Section 3 describes the proposed methods. Section 4 details the experimental setup and the produced results. Finally, Sect. 5 concludes the paper and provides directions for future research.

2   Related Work

In this section we present related work that utilize deep reinforcement learning and describe different methods proposed in the literature to combine learning from demonstrations and experience.

Recurrent neural networks can be used to generate sequences by considering the past generated samples. Such an approach is used in [1] to generate continuous handwriting. An extension to this approach is also proposed in [1] that allows the generated sequence to be conditioned on a sequence of input text characters. Clearly, such approaches can be very relevant to imitation learning if the actions can be formulated as a generated sequence conditioned on a sequence of observed states. Similarly sequence to sequence learning [19] has been gaining a lot of attention recently. However, for most applications, the entire input sequence is analysed before generating the out sequence, while autonomous agents are required to act in real time to every sensory input.

An LSTM based system is proposed in [11] to learn how to perform surgical procedures by controlling a robotic arm. The network is trained on demonstrations by a human expert. Although the static setting of the surgery allows for policies that replicate manually designed trajectories, this supervised learning approach provides better generalization.

The robocup simulator is a popular benchmark for intelligent learning methods as it shares many characteristics with real world applications. A cooperative defensive task is learned in [15] using demonstrations provided by two human players simultaneously. Several classifiers are used to learn from the demonstrations and the results are favourably compared to human performance and simple hand-coded agents. However, this approach employs high level strategies as the decision to be learned by the agent, such as "approach the ball" or "block attacker's path" which in turn need to be translated into low level actions through manual programming. High level actions also enable learning the task through evolutionary algorithms [14] as the solution space becomes smaller. Similarly, in [8,10,18] reinforcement learning is used to learn high level actions. It is noteworthy that each paper employs a different set of macro-actions; so each new macro-action has to be manually designed. Deep reinforcement learning is used in [5] to learn an offensive task from raw sensory data. The reinforcement learning policy is used to predict low-level parametrized actions and thus doesn't require manual policy design. However, the organic reward in this task (scoring a goal) is very sparse and requires performing long trajectories of low-level actions to reach this state. As reinforcement learning exploration fails to reach the environment's reward, this approach employs a manually engineered reward function that guides the agent to perform desired behaviours. This engineering requires substantial task knowledge and limits the general application of this approach.

3   Method

This section presents the proposed method for training agents to play soccer via deep imitation learning with memory. A recurrent neural network is trained solely from demonstrations and doesn't require any explicit tailored engineering. The proposed method minimizes the need for expert knowledge by utilizing the low level sensory features and learning a mapping to atomic parametrized actions. We start by describing the process of data collection and representation. Demonstrations are provided by a teacher that performs the task for a number of rounds. For each round the teacher attempts to score a goal; the round ends with a successful or unsuccessful attempt. A plethora of hand-crafted agents exist for the 'robocup soccer simulator' and can serve as the teacher to provide examples of effective behaviour. Existing agents are also available to control the opponent to provide a realistic setting for the demonstrations.

Each round is represented as a sequence of state-action pairs. For each frame t the state of the environment x_t is captured along with the action taken by the teacher y_t and are added to the sequence $S_i = x, y$. The state x_t represents low level information about the agent's surroundings and is captured from the agent's point of view using its simulated sensors. So all the information about the field and the objects and players in it are captured relative to the agent's position and status. The action y_t is chosen from a set of the low level parametric actions available to the agent. That is, the agent decides what move to perform from its list of actuators as well as one or more continuous values that serve as parameters for the selected actuator. Such atomic actions performed in a sequence construct a higher level behaviour that is usually identified and modelled manually in other studies. The captured sequences are used to construct the training dataset $D = S_1, S_2..S_n$ is used to train recurrent neural network.

The training set is used to train a deep recurrent neural network. The network consists of 3 stacked LSTM layers containing 100, 50 and 6 nodes respectively, followed by a reshaping layer to present the 6 output values for all samples in the batch to the loss function. The loss function calculates the error for the predictions of the entire batch rather than the final prediction only. This is because unlike most RNN applications we are interested in producing accurate predictions at each frame rather than optimizing one prediction after reading the entire sequence. The LSTM layers are used to extract high level temporal features from the raw input and the context provided by the networks memory.

The LSTM layers utilize hyperbolic tangent (tanh) activation functions. The output nodes in the final layer correspond to parametrized actions and are used to predict continuous values for the 6 possible parameters for the agent's actuators. The output layer utilizes linear activations and a mean square error loss function is used, therefore the network behaves as a multivariate regressor. The actuator with the highest predicted parameter value is selected for execution by the agent. This method allows for prediction values for multiple parameters simultaneously. In many applications the output of the nodes in the final layer is not produced until the end of the sequence and is only fed into the next time step without being output as the network's prediction. In contrast, the proposed network does not read the entire sequence before producing a decision or generating an output sequence. Instead, at each instance of the input sequence the network predicts an output. Thus generating the output sequence step by step with the input, at each step utilizing all the information available up to this instance. By representing the demonstrations as sequences, this approach provides context for most of the samples facing the agent.

However, this makes the prediction dependent on the position of the sample in the sequence. For example if the agent starts performing the trained policy mid episode, the current frame will be treated as if it is at the beginning of the sequence even though it is not. To ensure the stationarity of the agent's policy, we train another network on a modified version of the training set D in which all the sequences $S_1, S_2..S_n$ are augmented into one list of samples. This list is subsequently segmented into segments of uniform length that serve as the new training sequences to be fed into the LSTM network. Figure 1 illustrates the segmentation of the artificial sequences. This arbitrary creation of sequences presents different states in different parts of the training sequences while maintaining a temporal dependency between the consecutive instances in a sequence. This approach is not expected to outperform training on fully structured sequences given that complete sequences are always presented to the agent during testing. However, it demonstrates that the proposed approach does not depend on reproducing entire training sequences and that utilizing memory in imitation learning is beneficial even if the beginning and end of the sequence are unknown.

Table 1 highlights the differences between the proposed method and other intelligent methods used for "Robocup". Most methods rely on manually engineering features and high-level actions which require significant task specific knowledge and engineering which does not allow for a general learning process. [5] uses deep learning to alleviate the need for engineering features and directly map raw features to low-level actions. However, designing dense reward functions to guide the agent require similar effort and produce similar results to manually engineering the policy. A change in the setting such as the number of players on the field requires designing new reward functions. This is in contrast with organic reward functions which are directly provided by the rules of the game. The proposed approach is general and only receives knowledge about the task from the demonstrations. Providing new demonstrations for changes in a task is considerably easier than designing reward functions or low-level policies to execute the high-level decisions made by the policy.

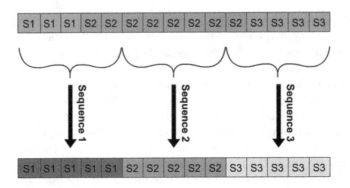

Fig. 1. Re-segmenting the demonstrated sequences into arbitrary sequences

Table 1. A comparison of machine learning approaches for rocbocup

Method	Learning	Features	Actions	Rewards
Jain et al. [8]	Reinforcement learning	Selected	High-level	Engineered
Raza et al. [15]	Supervised (various)	Engineered	High-level	N/A
Stone et al. [18]	Reinforcement learning	Engineered	High-level	Organic
Masson et al. [10]	Reinforcement learning	Selected	High-level parametrized	Engineered
Hausknecht et al. [5]	Reinforcement learning	Raw	Low-level parametrized	Heavily-engineered
Ours	Supervised (LSTM)	Raw	Low-level parametrized	N/A

The proposed LSTM network is compared to a multi-layer perceptron that doesn't have a memory and treats all frames as independent and identically distributed samples. In this case the sequences are augmented to create one training set, from which batches of samples are drawn. Keeping the sequence of samples without utilizing memory can be detrimental to training as the samples in training batches will be too similar and lack diversity. Therefore, when training the MLP, the entire dataset is shuffled before sampling the training batches to ensure that they contain diverse samples from a variety of situations. This is similar to the replay buffer approach used in [13] which is a key factor in the success of deep reinforcement learning. The architecture of the MLP consists of 3 fully connected layers containing 100, 50 and 6 nodes respectively. The first 2 layers utilize rectifier activation functions and the output layer uses a linear activation function.

Moreover, to evaluate the generalization of the proposed LSTM approach, it is compared to a data aggregation approach that is applied during MLP training. DAGGER [16] is a seminal data aggregation method that aims to enhance generalization in imitation learning by providing additional training samples based on the agent's initially trained policy. The agent is allowed to perform the task using the policy trained using the MLP. For each frame the teacher provides the optimal action for the state observed by the agent and a new training dataset is collected. The agent stochastically chooses to perform the teacher's instruction or the action predicted based on its current policy. The new samples are added to the original training set and used to train a new agent. The new samples show states that are likely to be visited by the agent according to its trained policy and thus improving generalization. This process is repeated iteratively and the new set of training instances are aggregated into the final training set.

4 Experiments

4.1 Robocup

The robocup soccer simulator [9] is a 2D simulator that allows for full soccer matches between 11 player teams. The simulator is a popular benchmark for artificial intelligence as it contains a number of real characteristics and challenges found in real applications such as a dynamic multi-agent environment and relative sensory information. A challenge closely related to this study is the fact that playing soccer requires performing long sequences of actions that depend on previous actions as well as actions from other agents. Because soccer is a familiar activity, this application provides extensive evaluation of the agents' performance; not only according to the well-established rules of soccer but also qualitatively analysing the agents' behaviour through 2D visualization. Implementation of the proposed methods is available at https://github.com/ ahmedsalaheldin/RoboCupLSTM.

4.2 Half-Field-Offence

This study is conducted on a simplified sub problem of soccer simulator called Half-field-offence (HFO) [4]. Over the years, researchers have used simplified versions of the game of soccer to create intelligent autonomous agents in a more controlled setting [17]. As the name suggests HFO takes place in half the soccer field and is only concerned with the task of offence. The round is initiated with the offensive player and the ball randomly placed in the half-field. The objective of the offence is to score in the opponents goal while the defence tries to intercept the ball. The round ends if a goal is scored or the defence captures the ball or the ball goes out of the half-field bounds or if a time-limit is reached. Full details of the available features and actions can be found in [4]. In our experiments we use demonstrations provided by the offensive agent to teach an intelligent agent to play the offensive role. A simple hand crafted agent is used as the teacher to facilitate communication with the server. More complex higher performing hand crafted agents can be used to provide demonstrations in the future. Figure 2 shows the visualization of the HFO environment.

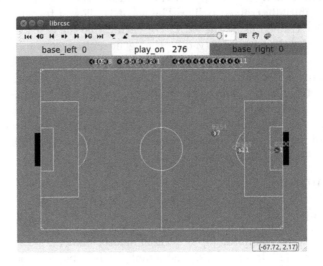

Fig. 2. Illustration of the Half-field-offence environment

4.3   Experimental Setup

The experimental evaluation compares between 4 learning methods. Firstly the proposed LSTM trained on the captured sequences. This method referred to as "LSTM episodic" as each episode or round of HFO makes up a training sequence. Secondly, "LSTM segmented", where the LSTM model is trained on the uniformly segmented sequences. The third method "MLP" trains a supervised multi-layer perception on the training set. And finally "MLP shuffled" is similar to "MLP" but shuffles the dataset before training the model. Moreover, data aggregation of 3 iterations is applied to the MLP based approaches. All models are trained on the same collected demonstrations consisting of 20000 samples. Data aggregation adds a further 5000 samples for each iteration. The sequence length used for training "LSTM segmented" is 80 samples. The models are trained offline for 1500 epochs and the trained networks are saved to be later used by the agent in real time.

We use a client that is decoupled from the learned models to connect to the simulator server so that the same client can be used to execute any learned policy. The client communicates with the simulator to receive the raw sensory data observed by the agent at each frame and send the decisions of the neural network to control the agent's actions. The models are evaluated on 1000 rounds of HFO. Each round can end in one of 4 outcomes. Firstly, a goal is scored, which is the best possible outcome, followed by the defence capturing the ball, then the ball going out of bounds and finally running out of time represents the poorest behaviour by the agent. Table 2 shows the percentage of each outcome achieved by the teacher used to provide the demonstrations; in 1000 rounds of playing.

Table 2. HFO results for the hand crafted teacher

Method	Goal	Defence	Bounds	Time
Teacher	44.37%	51.43%	4.19%	0%

4.4   Results

Firstly, the results comparing the proposed LSTM approach to imitation learning without memory are presented. Figure 3 shows the results for the 4 trained models "LSTM episodic", "LSTM segmented", "MLP" and "MLP shuffled". The results are shown for 1000 rounds of testing. The graph shows the percentage of rounds that resulted in the 4 possible outcomes: goal scored, captured by the defence, ball out of bounds and out of time. The percentage of goals scored is the most important measure as scoring is the primary objective of the task, however the other measures show the rest of the picture. The proposed method "LSTM episodic" has resulted in the highest percentage of goals, similar to the teacher's performance and outperforms networks without memory with statistical significance. "LSTM segmented" comes in second place also outperforming the MLP methods with statistical significance, demonstrating that utilizing memory is the contributing factor in the effectiveness of the learned policy, even if the beginning and end of the sequence are unknown.

The results also show that shuffling the training set resulted in significantly more goals. This corroborates the hypothesis that using dependent sequences of samples to train models without memory can be detrimental as the training batches lack diversity. The remaining measures show the robustness of utilized imitation learning methods with small percentages of unwanted outcomes ("out of bounds" and "out of time") especially the LSTM based methods. Qualitative analysis of the performance shows that running out of time is usually the result of the agent getting stuck and constantly performing the same action. As can be expected the teacher never produced this outcome and it is considered the poorest behaviour displayed by the imitating agents. Being stuck is an indication of ambiguity in the agent, and the extremely low percentage of this behaviour in the LSTM agents demonstrates that utilizing memory significantly improves the generalization of the learned policy in addition to its effectiveness.

Following, the results for data aggregation are presented. Figure 4 shows the results for using data aggregation on the MLP network, with and without shuffling the data. With "MLP shuffle", the entire data set is shuffled each iteration, so the training batches can contain samples from all the aggregated datasets. The graph shows the percentage of goals scored in 1000 rounds for supervised learning (using the original training set), and three iterations of data aggregation.

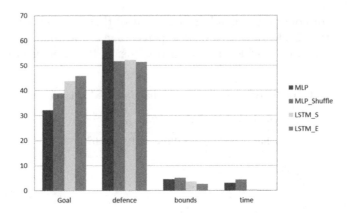

Fig. 3. Results for robocup half field offence. The outcomes presented in the graph are: Goal: The offensive agent scored a goal, Defence: The ball was captured by the defence, Bounds: The ball went out of bounds, Time: A time limit was reached before any of the other outcomes

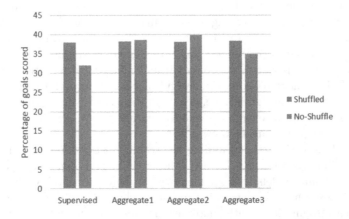

Fig. 4. Scoring percentage for MLP with data aggregation

The graph shows that for "MLP shuffle", there is no significant improvement in the percentage of scored goals. Without shuffling, we can see an improvement in scoring for the first two DAGGER iterations but this pattern doesn't hold for the third iteration. For both methods, the graph shows no consistent improvement in scoring with increasing the DAGGER iterations and in all cases doesn't reach the performance of the LSTM approaches.

Tables 3 and 4 shows the complete results for data aggregation with multilayer perceptrons, with and without shuffling respectively. The results show that aggregating new samples does not necessarily decrease the percentage of undesirable outcomes. For both approaches, there does not appear a pattern for decreasing the "out of time" percentages with increasing iterations of data aggregation.

Table 3. MLP data aggregation results with shuffling

Method	Goal	Defence	Bounds	Time
Supervised	38.83%	51.74%	5.02%	4.40%
Aggregate 1	38.17%	55.12%	4.46%	2.23%
Aggregate 2	38.01%	53.09%	5.68%	3.20%
Aggregate 3	38.31%	53.14%	2.98%	5.56%

Table 4. MLP data aggregation results without shuffling

Method	Goal	Defence	Bounds	Time
Supervised	32.19%	60.12%	4.56%	3.11%
Aggregate 1	38.59%	49.08%	4.27%	7.94%
Aggregate 2	39.81%	54.86%	3.78%	1.53%
Aggregate 3	34.89%	59.89%	3.67%	1.53%

In "MLP no-shuffle" where the scoring rate was substantially improved with the first iteration of data aggregation, we observe that this improvement is accompanied by a huge rise in the percentage of "out of time" rounds. This emphasises that data aggregation in this study doesn't provide a consistent improvement in the agent's performance. Although data aggregation utilizes more information, by sampling demonstrations from likely states, it fails to improve the generalization of the agent compared to the proposed LSTM approach.

5   Conclusion and Future Work

This paper proposes a novel imitation learning approach for learning from sequences in a dynamic environment. A demonstration is represented as an ordered sequence of state-action pairs. The states are represented by a feature vector of low level sensory information from the agent's perspective. The actions available to the agent are low level parametrized actions. A deep Long-short-term-memory network is used to learn a policy that retains a memory of past experiences and learns from the entire demonstrated trajectory of actions. The trained model uses memory to provide context to improve generalization and predicts an action at every frame in real-time. Results on a multi-agent soccer simulator show that learning from sequences using memory networks can significantly outperform learning from i.i.d. samples and reach comparable performance to the teacher. Using the memory to provide context when learning from sequences outperforms data aggregation methods for improving generalization and is much faster to train. Moreover, it is also shown that the proposed LSTM method can be stationary by training on sequences that are arbitrarily segmented from the demonstrations without a significant drop in performance. Experiments using multilayer perceptions show that if the model has no memory when learning from sequences, shuffling the training data can result in a significant improvement in performance as the samples in the training batches become more diverse. In the next step, we aim to use a number of high performing agents from the robocup competitions to provide the demonstrations and include more agents in the game.

References

1. Graves, A.: Generating sequences with recurrent neural networks. arXiv preprint arXiv:1308.0850 (2013)
2. Graves, A., Mohamed, A.R., Hinton, G.: Speech recognition with deep recurrent neural networks. In: 2013 IEEE International Conference on Acoustics, Speech and Signal Processing (ICASSP), pp. 6645–6649. IEEE (2013)
3. Graves, A., et al.: Supervised Sequence Labelling with Recurrent Neural Networks, vol. 385. Springer, Heidelberg (2012). https://doi.org/10.1007/978-3-642-24797-2
4. Hausknecht, M., Mupparaju, P., Subramanian, S., Kalyanakrishnan, S., Stone, P.: Half field offense: an environment for multiagent learning and ad hoc teamwork. In: AAMAS Adaptive Learning Agents (ALA) Workshop (2016)
5. Hausknecht, M., Stone, P.: Deep reinforcement learning in parameterized action space. arXiv preprint arXiv:1511.04143 (2015)
6. Hussein, A., Elyan, E., Gaber, M.M., Jayne, C.: Deep imitation learning for 3D navigation tasks. Neural Comput. Appl. 1–16 (2017)
7. Hussein, A., Gaber, M.M., Elyan, E., Jayne, C.: Imitation learning: a survey of learning methods. ACM Comput. Surv. 50(2), 21:1–21:35 (2017). https://doi.org/10.1145/3054912
8. Jain, D., Shah, M., Garg, B.K.: Watchdogs 2D soccer simulation
9. Kitano, H., Asada, M., Kuniyoshi, Y., Noda, I., Osawa, E.: RoboCup: the robot world cup initiative. In: Proceedings of the First International Conference on Autonomous Agents, AGENTS 1997, pp. 340–347. ACM, New York (1997). https://doi.org/10.1145/267658.267738
10. Masson, W., Ranchod, P., Konidaris, G.: Reinforcement learning with parameterized actions. arXiv preprint arXiv:1509.01644 (2015)
11. Mayer, H., Gomez, F., Wierstra, D., Nagy, I., Knoll, A., Schmidhuber, J.: A system for robotic heart surgery that learns to tie knots using recurrent neural networks. Adv. Robot. 22(13–14), 1521–1537 (2008)
12. Mikolov, T., Karafiát, M., Burget, L., Cernockỳ, J., Khudanpur, S.: Recurrent neural network based language model. In: Interspeech, vol. 2, p. 3 (2010)
13. Mnih, V., et al.: Playing atari with deep reinforcement learning. arXiv preprint arXiv:1312.5602 (2013)
14. Pietro, A.D., While, L., Barone, L.: Learning in RoboCup keep away using evolutionary algorithms. In: Proceedings of the 4th Annual Conference on Genetic and Evolutionary Computation, pp. 1065–1072. Morgan Kaufmann Publishers Inc. (2002)
15. Raza, S., Haider, S., Williams, M.A.: Teaching coordinated strategies to soccer robots via imitation. In: 2012 IEEE International Conference on Robotics and Biomimetics (ROBIO), pp. 1434–1439. IEEE (2012)

16. Ross, S., Gordon, G.J., Bagnell, J.A.: A reduction of imitation learning and structured prediction to no-regret online learning. arXiv preprint arXiv:1011.0686 (2010)
17. Stone, P., Kuhlmann, G., Taylor, M.E., Liu, Y.: Keepaway soccer: from machine learning testbed to benchmark. In: Bredenfeld, A., Jacoff, A., Noda, I., Takahashi, Y. (eds.) RoboCup 2005. LNCS (LNAI), vol. 4020, pp. 93–105. Springer, Heidelberg (2006). https://doi.org/10.1007/11780519_9
18. Stone, P., Sutton, R.S., Kuhlmann, G.: Reinforcement learning for robocup soccer keepaway. Adapt. Behav. **13**(3), 165–188 (2005)
19. Sutskever, I., Vinyals, O., Le, Q.V.: Sequence to sequence learning with neural networks. In: Advances in Neural Information Processing Systems, pp. 3104–3112 (2014)

Toward Video Tampering Exposure: Inferring Compression Parameters from Pixels

Pamela Johnston[1($\boxtimes$)], Eyad Elyan[1], and Chrisina Jayne[2]

[1] Robert Gordon University, Aberdeen, UK
1609098@rgu.ac.uk
[2] Oxford-Brookes University, Oxford, UK

Abstract. Video tampering detection remains an open problem in the field of digital media forensics. Some existing methods focus on recompression detection because any changes made to the pixels of a video will require recompression of the complete stream. Recompression can be ascertained whenever there is a mismatch between compression parameters encoded in the syntax elements of the compressed bitstream and those derived from the pixels themselves. However, deriving compression parameters directly and solely from the pixels is not trivial. In this paper we propose a new method to estimate the H.264/AVC quantisation parameter (QP) in frame patches from raw pixels using Convolutional Neural Networks (CNN) and class composition. Extensive experiments show that QP of key-frames can be estimated using CNN. Results also show that accuracy drops for predicted frames. These results open new, interesting research directions in the domain of video tampering/forgery detection.

Keywords: CNN · Compression · Video tampering detection

1 Introduction

In the age of fake news and falsified video, the detection of video tampering is becoming an increasingly important area of research. Machine learning techniques [1,2] can be used to alter video content by changing faces or weather conditions, yet detection of such tampering remains an open field. Detection methods can be active or passive [3,4], but, since many existing videos are unprepared for active tampering detection, passive detection methods are more relevant. Passive tampering detection can be categorised into recompression, region tampering and inter-frame forgery [3]. Region tampering includes copy-move attacks where the copied region can come from the same frame in the video, similar to image copy-move [5] or from a different spatio-temporal region in the same video [6]. Splicing and inpainting are variations on region tampering. Inter-frame forgery is where an integer number of frames is added, deleted or shuffled. Regardless of the editing method, however, any tampering at the pixel level of a

© Springer Nature Switzerland AG 2018
E. Pimenidis and C. Jayne (Eds.): EANN 2018, CCIS 893, pp. 44–57, 2018.
https://doi.org/10.1007/978-3-319-98204-5_4

compressed video requires recompression of the video bitstream [7,8], so of these three methods, recompression detection is the most versatile.

Video compression is prevalent in digital society. The vast majority of online video has been compressed using lossy formats such as H.264/AVC [9] or MPEG2 [10]. These formats have been designed with the human visual system in mind and the effects of compression remain largely invisible to human eyes. It has been shown that compression does impact classification performance of convolutional neural network (CNN) classifiers [11] and pre-existing compression in original source images may even have caused these effects to be understated. If CNN classifiers are passively affected by compression, it is reasonable to use them to actively detect the level of compression directly from pixels. Moreover, any method of measuring compression could be utilised to create an ensemble CNN classifier which could maximise accuracy while accounting for recompression. Accurate QP estimation could be used to enhance the performance of classifiers across differing quality levels.

An intuitive indication of recompression is where the Quantisation Parameter (QP) encoded within the bitstream fails to match the value estimated from the pixels. This is most obvious to human eyes when the bitrate and syntax elements of the bitstream imply high quality video data but the pixel content exhibits visible compression artifacts such as blockiness. Accurate QP estimation from pixels would allow recompression detection which would aid tampering detection. The human visual system cannot distinguish between close QP levels, however and objective methods of measuring QP from pixels are required. An ideal QP estimator would also operate accurately over small patches to enable localisation of tampered regions because this is an advancing area of research [4,12]. For singly compressed frames, estimated QP can be verified by encoded bitstream syntax elements. In multiply compressed video, there will be mismatches between estimated QP and syntax elements, and differing QP patterns may be detected over spatially or temporally tampered regions.

This work takes a step towards utilising compression parameters derived directly from the pixels themselves. We show CNNs can be trained to estimate QP for stand alone key frame patches with reasonable accuracy. Original datasets are synthesised from uncompressed sources and used to train CNNs to identify the QP used to encode the data. We train a CNN to estimate the quantisation parameter of a pixel patch singly encoded using H.264/AVC. The accuracy of our model is examined and contributory factors to errors, including the reasons for lower accuracy on predicted frames, are explored. Class composition is also used to improve accuracy in predicted frames. Unlike [13], where decomposition of a original dataset classes into *smaller* subclasses improved accuracy in random forest classification, we combine adjacent classes into *larger* superclasses. We find that training a CNN on superclasses improves accuracy. We explain how our model works through examination of the network weights.

2   Background and Related Work

The human visual system is adequate to detect some compression effects and can quantify "no reference" image and video quality [14,15]. The source of video compression visual effects can be found by examining transformations used in compression standards. A video sequence comprises key or intra frames, which provide access points into the sequence, and predicted frames which rely on data from previously encoded frames. In H.264/AVC and MPEG-2, frames are divided into "macroblocks": blocks of 16 x 16 pixels. For non-predicted data, the pixel data itself is transformed into the frequency domain using Discrete Cosine Transforms (DCT), quantised and variable length encoded for transmission. For predicted data, a suitable patch of reference pixels is located, then the *difference* between current and reference data is transformed, quantised and encoded. Quantisation is performed as in Eq. 1 where δ is DCT coefficients of a macroblock or residual, C is the compressed coefficients and Q_s represents the quantisation step as indexed by the quantisation parameter [16].

$$C = round(\frac{\delta}{Q_s}) \tag{1}$$

Higher QP indexes larger Q_s and means more frequency coefficients are filtered out entirely. An increase in QP often manifests visually as an increased "blockiness"; that is, discrete regions of macroblocks consisting single or few frequency coefficients. Most often, low frequencies have higher signal amplitudes, so sharp edges persist while textures are reduced. In key frames, macroblock edges align uniformly within the frame. This visual effect was more apparent in earlier video compression standards [10] where non-integer DCTs forced regular inclusion of key frames. Periodic key frames limited drift between encoder and decoder but were visible as a pulse in the sequence as accumulated rounding errors were reset by the key frame. The integer transforms introduced in H.264/AVC [9] reduced the role of key frames to access points in the bitstream and consequently reduced the visible pulse in video sequences. HEVC [17] defines other techniques to reduce visible compression artifacts but is yet to be fully adopted. H.264/AVC is more common in the wild. Compression artifacts are not restricted to artificial block edges, however, and can also manifest as a lack of specific frequency detail or as banding in areas of smooth colour/intensity transition.

Traditional methods of recompression detection rely on the identification of patterns in frequency domain bitstream syntax elements. The authors of [18] rely on Benford's distribution of DCT coefficients and support vector machines to detect double compression of intra frames. In [19] multiple compression is detected in H.264/AVC encoded videos but the compression modes are heavily restricted and the methods do not differentiate between QP that are less than two steps apart.

As part of an investigation into using deep neural networks to determine image quality, Bosse et al. [14] developed a method to estimate QP of HEVC frames directly from pixels. They achieved accurate results for average QP estimation over a complete frame using a patch-wise technique and dataset synthesised from UCID [20]. The method was applied to intra (key) frames only. QP estimation was framed as a regression problem and the dataset used to train the network contained labelled patches compressed with all possible QPs. Although the averaged QP prediction for a complete frame was accurate, a heatmap showing individual patch contributions displayed great variation between patches. If QP estimation is to be successful as a region-tampering detector, it should be as accurate as possible over small regions. Moreover, a QP estimator for video must also handle *predicted* frames.

This work examines QP estimation in the context of patches taken from H.264/AVC video sequences. H.264/AVC is currently one of the most popular video compression standards and is used on YouTube, broadcast video and public datasets. A CNN is trained to classify frame patches from a video sequence using their quantisation parameters as labels. Unlike [14], we also investigate predicted frames in a video sequence.

3 Datasets and CNNs for Quantisation Estimation

3.1 Datasets

When examining the effects of compression, is vital to start with unprocessed data. Standard YUV 4:2:0 sequences from xiph.org are commonly used for video compression quality analysis[1]. Strictly speaking, YUV 4:2:0 is a compressed format due to reduced resolution of the colour channels but it is widely used in video compression. Uncompressed YUV 4:4:4, is not as popular. The sequences from xiph.org come in various dimensions and cover a wide variety of subjects from studio-shot sequences to outdoor scenes. All sequences are single camera, continuous scenes. Camera motion varies between sequences but frames from a single sequence will be correlated.

A large amount of data is required to train a neural network and uncorrelated data will produce a more generalised network. It is possible to use still image data as single frame sequences when focussing on spatial compression artifacts and excluding temporal compression. For this purpose, the images of UCID [20] were used. UCID consists of uncompressed images which are either 512×384 pixels or 384×512 pixels and cover a wide variety of subject matter. All are natural scenes and taken with the same camera. Of the original reported 1338 images in the dataset, only 882 were available for download[2]. Using a dataset of single images is not ideal since predicted frames cannot be examined. However it allows for a greater variety of pixel combinations in a smaller dataset because individual images are uncorrelated. Each image from UCID was regarded as a single frame video sequence and encoded accordingly as an intra frame.

---

[1] Available from Derf's Media Collection: https://media.xiph.org/video/derf/.
[2] UCID images from http://jasoncantarella.com/downloads/ucid.v2.tar.gz.

Table 1. A summary of original datasets

Name	Source	Length	Dimensions	Key frames
CIFvid	xiph.org	18 videos	352 x 288	1/250
CIFintra	xiph.org	18 videos	352 x 288	All
AllVid	xiph.org	44 videos	176 x 144 to 1920 x 1080	1/250
AllIntra	xiph.org	44 videos	176 x 144 to 1920 x 1080	All
UCID	UCID [20]	882 single frames	512 x 384 or 384 x 512 pixels	All

Table 1 gives a summary of the original datasets. Each video sequence was compressed using the open source H.264/AVC encoder x264 and one of a range of constant QP levels using variable bitrate mode. Constant quantisation parameters were selected with an even distribution: QP = [0, 7, 14, 21, 28, 35, 42, 49]. Constant bitrate rate control, psychovisual options and deblocking filter were turned off. For datasets containing predicted frames, the key frame interval was 250. Patches were then extracted from the decoded YUV4:2:0 sequences. Patches were converted from YUV4:2:0 to YUV4:4:4, where the Y-channel represents intensity and U and V channels are colour. Table 2 summarises synthesised datasets. A large temporal stride was used to limit correlation between patches. Consecutive frames are similar to each other and training a neural network with a correlated dataset will cause overfitting. Each patch was labelled with its quantisation parameter. All datasets were prepared in advance of network training and the original video sequences were split into train and test sets prior to compression and patch sampling to prevent data leakage[3].

Table 2. A summary of synthesised datasets

Name	Source	Patch size	Spatial stride	Temp. stride	Train patches	Test patches
AllVid_80	AllVid	80	80(train); 40(test)	40	156592	8400
AllIntra_80	AllVid	80	80(train); 40(test)	40	156592	8400
UCID_80	UCID	80	80	1	131904	53480
CIFvid_80	CIFvid	80	48	30	79920	7920
AllVid_32	AllVid	32	80(train); 40(test)	40	191776	13872
AllIntra_32	AllVid	32	80(train); 40(test)	40	191776	13872
UCID_32	UCID	32	80	1	183320	26320
UCID_32_large	UCID	32	32	1	974528	140512
CIFvid_32	CIFvid	32	32	60	118976	12672
CIFintra_32	CIFvid	32	32	60	118976	12672

---

[3] Training sequences: akiyo, bridge-close, bridge-far, carphone, claire, coastguard, foreman, hall, highway, mobile, mother-daughter, paris, silent, stefan, waterfall, old_town_cross, crowd_run, ducks_take_off, in_to_tree, mobcal, old_town_cross, parkrun, shields. Test sequences: bus, flower, news, tempete.

Two different patch sizes were selected to investigate which aspects of compression were important to CNNs. Block edge artifacts in intra frames will present themselves at macroblock (and subblock) boundaries. Therefore, any patch size larger than 16×16 will capture block edge artifacts. Following [14], a small patch size of 32×32 was selected. A larger patch size of 80×80 pixels was also used. When aligned with the macroblock grid, 80×80 pixels covers 5×5 complete macroblocks. A larger patch size allows for more context and image features within the patch to contribute towards QP estimation. Spatial strides were selected so that there was no patch overlap in the training set, although patches taken from the same video sequence would exhibit some correlation.

3.2 Network Architectures

For the purposes of this paper, three simple network architectures (NAs) were examined, summarised in Table 3. Image patches were format YUV 4:4:4, rescaled to values between 0 and 1 and whitened. In order to preserve compression artifacts in situ, no further data augmentation was used. Batch size was 128 patches. Unless otherwise noted in the results, NAs 1 and 2 were implemented with stride = 2 for all convolutional and pooling layers.

Network architectures were designed with compression artifacts in mind. H.264/AVC uses a minimum DCT block size of 4×4 pixels so a 4×4 kernel aligns to this. Using an even-sized kernel is unusual but not without precedent [21]. A stride of 2 allows sufficient overlap to encounter artifacts while reducing the number of network parameters. Networks were trained and tested multiple times and average accuracy and confusion matrix values taken for the results.

Table 3. A summary of network architectures

Name	Layers
NA 1	conv4x4-64, pool3x3, norm, conv4x4-64, norm, pool3x3, fc-384, fc-192, softmax
NA 2	conv5x5-64, pool3x3, norm, conv5x5-64, norm, pool3x3, fc-384, fc-192, softmax
NA 3 [14]	conv3x3-32, conv3x3-32, pool2x2, conv3x3-64, conv3x3-64, pool2x2, conv3x3-128, conv3x3-128, pool2x2, conv3x3-256, conv3x3-256, pool2x2, conv3x3-512, conv3x3-512, pool2x2, fc-512, softmax

3.3 Estimating Quantisation Accuracy

The quantisation parameter (QP) in H.264/AVC can be expressed as:

$$0 \leq QP \leq 52, \; QP \in \mathbb{R} \tag{2}$$

QP relates directly to Q_s in Eq. 1. Patches with similar QP labels exhibit similar compression features, and confusion matrices produced by the network reflected this. Two different QPs might have very similar effects on a given patch, depending on the patch content. An example of this is a whole patch of solid colour, which transforms to a single high amplitude, low frequency coefficient which is non-zero on quantisation. Such an extreme example is unlikely in natural scenes but it demonstrates how applying close QPs might result in identical patches with different labels. Therefore, QP has been sampled at [0, 7, 14, 21, 28, 35, 42, 49] in the synthesised datasets. Using all possible QP would also generate an extremely large dataset and increase model training times. Using a range of sampled QP, the confusion matrices produced by the model can be examined and super-classes composed to estimate accuracy.

Unlike the work presented in [13], where data was decomposed based on discrete classes, in this paper, we combine each two adjacent classes into one, assuming that the change in pixels is not significant within adjacent QP. Figure 1 gives a visual demonstration of how this can be applied to a confusion matrix. In a confusion matrix, predicted labels from the network are tabulated against ground truth class labels. The overall accuracy of the network is given as the sum of the diagonal elements of the confusion matrix divided by the sum of all the elements (Fig. 1a). That is, the *correctly predicted* elements divided by all the elements. If some degree of error is permitted in the confusion matrix, adjacent classes can be accepted as "correct" and a new accuracy shown $\hat{p}$ (Fig. 1b) can be calculated by combining adjacent classes thus:

$$\hat{p} = \frac{1}{M}\{\sum_{i=0}^{m} a_{i,i} + \sum_{i=1}^{m} a_{i-1,i} + \sum_{i=1}^{m} a_{i+1,i} + \sum_{i=1}^{m} a_{i,i-1} + \sum_{i=1}^{m} a_{i,i+1}\} \quad (3)$$

Although Eq. 3 combines adjacent classes well in theory, testing how well it represents a model where classes are combined *before* training would involve assigning different labels to identical data. Instead, we compose super classes according to Eq. 4 and Fig. 1c.

$$\hat{p} = \frac{1}{M}\{\sum_{i=0}^{m/2} a_{2i,2i} + \sum_{i=0}^{m/2} a_{2i+1,2i} + \sum_{i=0}^{m/2} a_{2i,2i+1} + \sum_{i=0}^{m/2} a_{2i+1,2i+1}\} \quad (4)$$

Using Eq. 4, labels can be unambiguously combined and a network trained on these labels. Equation 4 can also be extended to create even larger super-classes allowing for ever greater error.

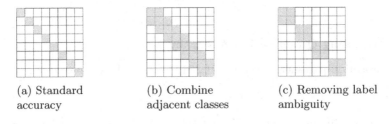

(a) Standard accuracy

(b) Combine adjacent classes

(c) Removing label ambiguity

Fig. 1. Different class compositions in a confusion matrix

4   Evaluation and Discussion

The initial experiment using CIFVid_80 achieved only 36.25% accuracy. Following [14], patch size was reduced and UCID was introduced, creating two new datasets: CIFVid_32 and UCID_32_large. The results in Table 4 show that smaller patch size did not improve accuracy, but training on intra frames only did. Halving the network stride parameter helped, but was still worse than using a larger patch size. Networks trained on UCID_32_large achieved accuracy of over 58% when tested with UCID_32_large test data, but approximately half that when tested with CIFVid_32.

Table 4. Initial results: accuracy for patch size 32 (network 3 failed to train).

Network	Tested on	CIFVid_32 trained	UCID_32_large trained
1	CIFVid_32	27.30	33.14
1	UCID_32_large	31.00	58.55
2	CIFVid_32	26.69	30.35
2	UCID_32_large	32.23	59.28
2	CIFintra_32	27.30	54.14
2 (stride = 1)	CIFVid_32	25.34	33.67
2 (stride = 1)	UCID_32_large	28.46	66.88

CIFintra_32, comprising all key frames (Table 2) answered the question of why learning from UCID_32_large did not translate well to CIFVid_32. Patches in CIFVid_32 and CIFintra_32 come from exactly the same points in the video sequences, so their content is strongly visually correlated. Only the underlying compression modes differ. The best performing network architecture was re-tested with CIFintra_32. The accuracy on CIFintra_32 using the network trained on UCID_32 showed good improvement over testing on CIFVid_32 (54.14% vs 30.35%).

CIFVid_32, CIFintra_32 and UCID_32_large were mismatched in terms of patch quantity within the training sets. To investigate whether this accounted for some of the differences in accuracy, AllVid_32, AllIntra_32 and UCID_32 were created. Table 5 shows the accuracy achieved on each combination. The larger training set UCID_32_large improved accuracy by an average 7.9% on the UCID_32 test set but only average 2.34% on AllVid_32. The addition of extra training video patches when increasing CIFVid_32 to AllVid_32 did not increase accuracy. From these differences in performance, it can be concluded that the UCID-based datasets were less correlated than those derived from video sequences leading to more feature coverage and more generalisable networks. In Table 5 intra-only trained networks still out-performed those trained on AllVid_32, except in the case of AllIntra_32 versus AllVid_32. Given that AllVid_32 and AllIntra_32 contain visually similar pixel patches, this implies

Table 5. Cross evaluation on similar sized datasets. Accuracy for patch size 32 (network 3 failed to train) and for patch size 80

| Network | Patch size 32 | | | | Patch size 80 | | | |
	Tested on	AllVid_32 trained	AllIntra_32 trained	UCID_32 trained	Tested on	AllVid_80 trained	AllIntra_80 trained	UCID_80 trained
1	AllVid_32	26.72	24.68	31.34	AllVid_80	34.93	26.02	36.72
1	AllIntra_32	27.75	33.75	44.14	AllIntra_80	34.11	37.94	63.40
1	UCID_32	33.74	41.56	50.96	UCID_80	41.28	47.46	72.75
2	AllVid_32	26.78	25.38	31.79	AllVid_80	35.27	29.39	37.28
2	AllIntra_32	27.07	33.16	45.05	AllIntra_80	34.93	46.54	62.50
2	UCID_32	33.83	41.25	51.07	UCID_80	42.04	56.66	71.65
3	-	-	-	-	AllVid_80	29.97	24.91	29.35
3	-	-	-	-	AllIntra_80	29.94	42.67	55.95
3	-	-	-	-	UCID_80	38.99	53.81	61.17

that the network trained on AllVid_32 has learned some features distinct to predicted frames that do not translate to key frames. This pattern was repeated with patch size 80 x 80.

Larger patch size datasets AllVid_80, AllIntra_80 and UCID_80 were generated and used to train and test different network architectures. Table 5 shows the results. Comparing results for AllVid_80 with CIFVid_80 and AllVid_32 (35.27%, 36.25%, 26.78%, respectively) show that increasing the number of patches did not improve accuracy but increasing the patch size did. Although networks trained and tested on UCID_80 achieved good accuracy (72.75%, Table 5), the learning did not transfer to AllVid_80 (36.72%). It did, however, translate to AllIntra_80 (63.40%). From this, it can be deduced that QP can be successfully estimated directly from the pixels of *key* frames but does not translate well to *predicted frames*. Networks trained on predicted frame patches achieve lower accuracy than that those trained on key frame patches. Moreover, the accuracy of networks trained on UCID_80 is higher than those trained on AllIntra_80. This can be partly attributed to weaker correlation in UCID_80 image patches.

Overall, NA 3, the deepest network, had the lowest accuracy. The limited size of the datasets may have contributed to this, but it is more likely that the depth of the network did not help with compression features. Compression features in H.264/AVC are related to the size of the transforms used in the codec and these vary from 4 x 4 to 16 x 16 pixels. It can be deduced that though deeper networks go some way to accounting for differences in scale in traditional object classification, this is largely unnecessary when examining compression.

4.1 Relaxing the Problem

Although the overall accuracy achieved on a network trained on predicted frame patches from AllVid_80 was low, the confusion matrix implied a reasonable error rate. Figure 2a shows the confusion matrix for NA 2 trained/tested on AllVid_80. Average accuracy was 35.27%. With class labels combined as in Eq. 4, the accuracy estimated from the confusion matrix is 54.65%. A network trained on the reduced label dataset yields a comparable accuracy of 56.25%. Therefore, the results obtained from calculations on the confusion matrix after training are comparable to networks trained specifically on these super classes. Table 6 shows that this is true across all patch size 80 datasets. Composing super classes from adjacent classes prior to training reduces the number of labels and slightly enhances generalisation in the network, yielding slightly higher results from video-based datasets with correlation between video patches. The same pattern was repeated with other architectures, though results are omitted for space considerations. QP in predicted frames can be estimated to within ± 7 (one class) with more than 54% accuracy. Higher quality frames are more challenging and this may be attributed to the larger range of frequencies available in uncompressed data. CNN models cannot distinguish between frequencies removed by compression and those simply absent in the source data.

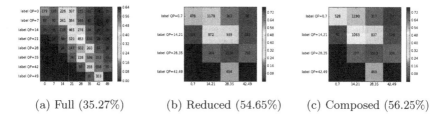

(a) Full (35.27%)   (b) Reduced (54.65%)   (c) Composed (56.25%)

Fig. 2. Confusion matrices for NA 2 trained/tested on AllVid_80 (overall accuracy for a single network)

Table 6. Accuracy for patch size 80 (NA 2): composition after/before training

Tested on	AllVid_80 trained	AllIntra_80 trained	UCID_80 trained
AllVid_80	54.65/56.25	52.99/54.08	60.70/59.55
AllIntra_80	55.24/56.51	66.62/67.55	78.81/78.47
UCID_80	66.81/67.94	78.58/79.57	88.43/88.41

The shape of the confusion matrices (Fig. 2) gives insight into the model's learning. Confusion matrices across all architectures and datasets demonstrated similar shapes where the bottom left corner approached zero. Patches of high QP were seldom misclassified as low QP. In contrast, the top right corner of the confusion matrix, although it displays lower numbers than the diagonal portion, does not always approach zero. This pattern suggests that the model has learned something about the frequency domain. Natural images contain a large variety of frequencies, however quantisation in the frequency domain selectively reduces these (as in Eq. 1). Weaker (low amplitude) frequency components are quantised to zero and thus filtered out of an image, leaving behind only dominant frequencies. For natural scenes, where lower frequencies tend to dominate, quantisation applied in video compression is effectively a low pass filter. This explains the "blocky" appearance of compressed video. Low amplitude, high frequency components aid smooth colour or intensity transition and removing these components leads to sharper colour transitions at macroblock edges. The detection of high frequency components within macroblocks therefore indicates lower QP. Unfortunately, the converse is not necessarily true. An absence of high frequencies does not indicate high QP, since some images naturally lack high frequency components.

(a) First layer filters   (b) 4x4 DCT

Fig. 3. Some of the first layer filters display visual similarity to a spatial representation of the 4 x 4 DCT transform used in H.264/AVC

4.2   First Layer Filters

A visual examination of the first layer filters confirms the presence of frequency features. Figure 3b shows a spatial representation of the 4 x 4 Discrete Cosine Transform used in H.264/AVC. These are pixels that result from an inverse DCT on a 4 x 4 coefficient matrix where only one coefficient is non-zero. Figure 3a shows selected first layer filters from NA 2 trained on AllIntra_32. Visual similarities are obvious. The CNN uses some first layer filters to infer frequency information directly from the pixels. Statistical analysis of frequency was also used in [19] to detect multiple compression.

4.3   Whole Image Heat Map

Figure 4 shows classification results for 80 x 80 patches of a key frame from the sequence "flowers" at QP 0 and 35. A plain black border was added added after compression to allow classification of 80 x 80 patches centred on every 16 x 16 macroblock. The heatmaps all show misclassification along the top row due to the black border. Comparing Figs. 4a and d, it is difficult for human eyes to differentiate between QP values, despite the large difference. The fine colour transition in the sky section is correctly classified by a network trained on UCID_80 as low QP. The network trained on AllVid_80 tends to overestimate sky QP and perform better on the colourful flower section. At moderate QP, the model trained on UCID_80 performs well. The AllVid_80 trained network achieves a reasonable mean over the whole image but underestimates QP in busy areas and overestimates in areas with fewer sharp edges. Mode 28 in Fig. 4f shows how the model confuses adjacent labels and validates the use of superclasses.

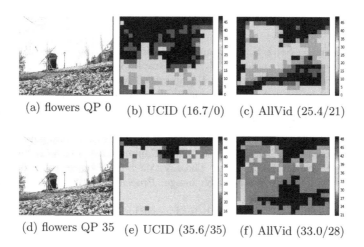

(a) flowers QP 0   (b) UCID (16.7/0)   (c) AllVid (25.4/21)

(d) flowers QP 35   (e) UCID (35.6/35)   (f) AllVid (33.0/28)

Fig. 4. The heat maps showing the predicted quant for a frame from the sequence "flowers", NA 2 used (**mean/mode**)

5   Conclusions and Future Work

We have shown that the level of compression of small image patches can be estimated objectively by CNN. The accuracy of CNNs trained on intra frames is much higher than those trained on predicted frames. The experimental results also strongly suggest that compression features learned from still images (intra frames) alone do not transfer to predicted frames. Although a neural network can be trained to estimate the QP in key frame patches, the results for predicted frames were weaker. In predicted frames, quantisation is applied to the residual difference between predicted and actual pixels. CNN compression estimation may be improved by using residuals from the compressed bitstream rather than reconstructed pixels. It may also be necessary to first identify intra macroblocks within an image in order to gain an estimate of accuracy in QP estimation.

Larger patch sizes yield higher precision but further investigation will clarify whether an optimum patch size exists. Smaller patch sizes are desirable if the model is to serve as accurate tampering detection. The features learned in the first layer of CNNs strongly resemble patterns of DCT coefficients, so QP estimation may be improved by initialising some of the first layer weights with DCT patterns.

Compression is no longer an irreversible "black box". Although recompression destroys information about original compression mechanisms in the compressed bitstream, tell-tale signs in the pixels can be used to estimate original levels and this information could be used to detect video tampering, including splicing and multiple compression.

References

1. Suwajanakorn, S., Seitz, S.M., Kemelmacher-Shlizerman, I.: Synthesizing obama: learning lip sync from audio. ACM Trans. Graph. **36**, 95 (2017)
2. Liu, M.-Y., Breuel, T., Kautz, J.: Unsupervised image-to-image translation networks. In: Advances in Neural Information Processing Systems (2017)
3. Sitara, K., Mehtre, B.M.: Digital video tampering detection: an overview of passive techniques. Digit. Invest. **18**, 8–22 (2016)
4. Qureshi, M.A., Deriche, M.: A bibliography of pixel-based blind image forgery detection techniques. Signal Process. Image Commun. **39**, 46–74 (2015)
5. Chauhan, D., Kasat, D., Jain, S., Thakare, V.: Survey on keypoint based copy-move forgery detection methods on image. Procedia Comput. Sci. **85**, 206–212 (2016)
6. Pandey, R.C., Singh, S.K., Shukla, K.K.: Passive forensics in image and video using noise features: a review. Digit. Invest. **19**, 1–28 (2016)
7. Singh, R.D., Aggarwal, N.: Video content authentication techniques: a comprehensive survey. Multimed. Syst. **24**, 211–240 (2017)
8. Ravi, H., Subramanyam, A.V., Gupta, G., Kumar, B.A.: Compression noise based video forgery detection. In: IEEE International Conference on Image Processing. IEEE (2014)
9. ITU-T. H.264: Advanced video coding for generic audiovisual services. ITU-T (2016)
10. ITU-T. H.262: Information technology - Generic coding of moving pictures and associated audio information: Video. ITU-T (2012)
11. Dodge, S., Karam, L.: Understanding how image quality affects deep neural networks. In: International Conference on Quality of Multimedia Experience. IEEE (2016)
12. Lin, C.-S., Tsay, J.-J.: A passive approach for effective detection and localization of region-level video forgery with spatio-temporal coherence analysis. Digit. Invest. **11**, 120–140 (2014)
13. Elyan, E., Gaber, M.M.: A fine-grained random forests using class decomposition: an application to medical diagnosis. Neural Comput. Appl. **27**, 2279–2288 (2016)
14. Bosse, S., Maniry, D., Wiegand, T., Samek, W.: A deep neural network for image quality assessment. In: IEEE International Conference on Image Processing. IEEE (2016)
15. Chen, Y.-J., Lin, Y.-J., Hsieh, S.-L.: Analysis of video quality variation with different bit rates of H. 264 compression. J. Comput. Commun. **4**, 32 (2016)
16. Richardson, I.E.: The H. 264 Advanced Video Compression Standard. Wiley, Hoboken (2011)
17. ITU-T. H.265: High efficiency video coding. ITU-T (2016)
18. Sun, T., Wang, W., Jiang, X.: Exposing video forgeries by detecting mpeg double compression. In: IEEE International Conference on Acoustics, Speech and Signal Processing. IEEE (2012)
19. Milani, S., Bestagini, P., Tagliasacchi, M., Tubaro, S.: Multiple compression detection for video sequences. In: IEEE International Workshop on Multimedia Signal Processing. IEEE (2012)

20. Schaefer, G., Stich, M.: Ucid: an uncompressed color image database. In: Storage and Retrieval Methods and Applications for Multimedia. International Society for Optics and Photonics (2003)
21. Bondi, L., Lameri, S., Güera, D., Bestagini, P., Delp, E.J., Tubaro, S.: Tampering detection and localization through clustering of camera-based CNN features. In: CVPR Workshops (2017)

RR-FCN: Rotational Region-Based Fully Convolutional Networks for Object Detection

Dingqian Zhang[1,3(✉)] ⓘ, Hui Zhang[2], Haichang Li[1], and Xiaohui Hu[1]

[1] Institute of Software Chinese Academy of Sciences, Beijing, China
{haichang,hxh}@iscas.ac.cn
[2] Beijing IrisKing Co., Ltd., Beijing, China
zhanghui@irisking.com
[3] University of Chinese Academy of Sciences, Beijing, China
zhangdingqian15@mails.ucas.edu.cn

Abstract. In this paper, we present rotational region-based fully convolutional networks (RR-FCN) for object detection. In contrast to previous detectors that do not consider rotation, our region-based detector incorporates rotational invariance into networks efficiently and generate more appropriate features according to the rotation angle. Specifically, we propose component-sensitive feature maps, rotational RoI pooling and interceptive back propagation which make RR-FCN learn rotation situations without extra supervision information. Using the 101-layer ResNet model, our method achieves state-of-the-art detection accuracy on PASCAL VOC 2007 and 2012. Moreover, since the feature maps in our network are component-sensitive, RR-FCN can find out objects with various postures, even those appear rarely in the training set. So our RR-FCN has better performance in the real world.

Keywords: Object detection · Rotational invariance
Fully convolutional network

1 Introduction

Recently, a series of deep network object detection methods have been proposed, such as fast R-CNN [5], faster R-CNN [15] and R-FCN [2]. These methods divide deep networks into two parts with Region of Interest (RoI) Pooling layer [5]: (i) a CNN-based part for extracting features from the whole image, and (ii) a part that is associated with RoI for classifying each proposal generated by RoI pooling layer. Because the second part gets location-aware proposals, the networks have the ability to detect objects in different places. However, object detection also requires rotational invariance. Rotational invariance means if an object rotates around its geometrical center, the prediction should be the same.

This work is supported by the Natural Science Foundation of China (U1435220) (61503365).

ⓒ Springer Nature Switzerland AG 2018
E. Pimenidis and C. Jayne (Eds.): EANN 2018, CCIS 893, pp. 58–70, 2018.
https://doi.org/10.1007/978-3-319-98204-5_5

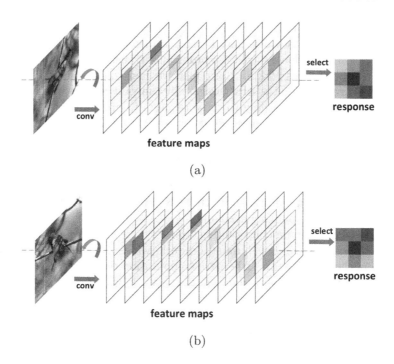

(a)

(b)

Fig. 1. Illustrations of our idea. Our network selects features from different locations on the feature maps, depending on the object under detecting. The images are the same in (a) and (b), but the rotation angles are different. RR-FCN gives them different feature selection methods. The colored bins mean they are selected.

Previous detectors train their parameters without considering rotational situations, which are called "*rotation tolerating*" in this paper. Instead of tolerating rotation, our network is aimed to understand rotation and select appropriate features. We call our process mode "*rotation handling*".

In this paper, we introduce a framework called Rotational Region-based Fully Convolutional Networks (RR-FCN) which can conveniently incorporate rotational invariance into deep networks for object detection. The idea is shown in Fig. 1. We use convolution operation to construct a group of component-sensitive feature maps. Each of these feature maps is sensitive to a specific component of an object, rather than a fixed part of region proposal. Then we use rotational RoI pooling proposed by us to select appropriate pieces of these feature maps to form the responses. Finally, we pick one of these responses as the output feature. Through this way, RR-FCN can deal rotation situations *without extra supervision information or parameters*. Moreover, since RR-FCN has component-sensitive feature maps, it can detect objects in various postures as long as the relative positions of components remain unchanged. So RR-FCN is more competitive in the real world.

In order to preserve the spatial information, we construct our network as a fully convolutional network. Recently, deep fully convolutional networks for computer vision become popular (e.g., [2,12,14]). Besides preserving spatial information of images, they also have fewer parameters and high computational speed. In [2], Dai et al. have solved the problem of translation variance fading in deep fully convolution network. In this paper, we figure out the detailed requests of translation invariance in the structure of fully convolutional networks. Further, we design our network according to these requests to preserve both translation variance and rotational invariance in our network.

Taking the 101-layer Residual Net (ResNet-101) [8] as our backbone and training it with interceptive back propagation developed by us, our RR-FCN has better robustness when the objects appear with rare postures. In simple terms, although people are barely reversed in the dataset, RR-FCN can still find an upside down person. Additionally, RR-FCN achieves state-of-the-art results. Our code will be made publicly available.

In conclusion, our main contributions are:

1. We propose a fully convolutional network, which can handle the rotation invariance of the targets with little extra computation time.

2. We preserve translation variance in the network by analysing and carefully designing the network structure.

3. We improve the robustness of the detection network. Our network can work normally even under the situations rarely appear in training set.

2   Related Works

In this section, we will discuss the previous works related to our work, covering the development of object detection and researches on rotational convolution networks.

2.1   Object Detection

Object detection networks can be divided into two kinds of frameworks: (i) detectors with region proposals [5,10,15], and (ii) detectors without region proposals [11,13,14]. In this paper, we focus on the first kind. Region-based object detection networks start using specialised pooling method from SPP-Net [7] (Spatial Pyramid Pooling) and Fast R-CNN [5] (Region of Interest pooling). Instead of resizing images into a fixed size, they pool the region proposals into a fixed size (e.g., 7 × 7). Then the fixed size outputs will be sent into various detectors to classify the related region proposals. R-FCN abandons fully-connected detector and shares the computation of the whole image for the first time. Without fully connected layers and deep RoI-wise sub-network, the detection precision becomes much lower than we expect. R-FCN uses position-sensitive RoI pooling to improve the detection result by solving translation variance problem in object detection, inspired by which we construct our networks. However, these methods all use the representation capacity and the powerful generalization ability to tolerate rotations of objects. We handle the rotational invariance problem in object detection which is also significantly important.

2.2 Rotational Convolution Networks

There is a research [1] indicating max pooling in CNN is beneficial to accept rotation. Max pooling ensures the output is constant when a few pixels on feature maps change their relative locations. However, it only works in the small region and ignores much useful information.

Several approaches have considered the rotation problems in convolutional networks. The most representative ones are Spatial Transformer Networks (STN) [9] and Deformable Convolutional Networks (DCN) [3]. STN adds a small network in parallel on the normal CNNs construction. This small network is capable of learning the affine transformation matrixes of images, in this way, it can accomplish translation, rotation, scale, etc. Although STN is successful in MNIST dataset, it fails to achieve desirable performance in object detection. This is due to the situation that detection datasets are relatively complex compared to MNIST. DCN also adds several convolution layers in parallel to guide the selection of sampling locations in convolutional operations. They both use bilinear interpolation to ensure back propagation can run during training. It is a success in semantic segmentation and improves the mAP in detection. However, DCN only selects the direction and distance of the expansion in convolution operations. Although its deformable position-sensitive RoI pooling is different from the previous pooling methods in the last sampling area, it still uses the relative position of the region proposal as a prior condition (e.g., the top-left part of a proposal). It cannot share rotational information across different feature maps. So the DCN focuses on deformation as its name indicates.

3 Our Approach

Figure 2 shows the basic architecture of our network. RR-FCN is an object detection network considering the rotation problem. Moreover, it incorporates rotational invariance into the network with no (or little) extra computation time. In this section, we describe the RR-FCN from three parts in detail.

3.1 Component-Sensitive Feature Maps and Rotational RoI Pooling

To make the feature maps in RR-FCN component sensitive, we develop a novel pooling method called rotational RoI pooling which pools feature maps clockwise. Figure 3 shows the relationship between component-sensitive feature maps and rotational RoI pooling. Each component-sensitive feature map is sensitive to a particular component of an object (e.g., head of a bird, leg of a dog). To distinguish different components, we divide every region proposal into $k \times k$ equal parts and each part has its corresponding feature maps. So we need to combine the $k \times k$ feature maps to get the entire detection result for one class. In our experiment, we make these components class-aware. That means each proposal has $k \times k \times (C + 1)$ feature maps. The C is the total number of categories in a dataset.

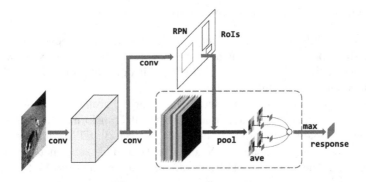

Fig. 2. The basic architecture of RR-FCN. We use different pooling methods with rotation information to get different responses from the same feature maps. We classify objects with the highest response. The proposed component-sensitive feature maps and rotational RoI pooling are marked with a red box. They can be used in parallel to obtain more specific information of rotation. Note that some back propagations in RR-FCN are not normal. (Color figure online)

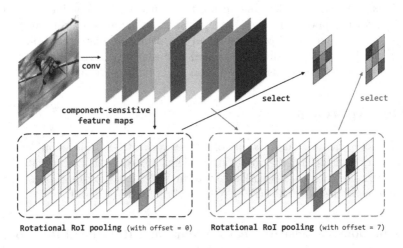

Fig. 3. Illustration of the component-sensitive feature maps and rotational RoI pooling. The feature maps may have multiple pooling methods with different offset f. So there are more than one output coming from a blank of feature maps. We take 0 and 7 offset for example.

Rotational RoI pooling layers pool component-sensitive feature maps into $k \times k$ bins with given rotation offset f which means the angle the features need to be rotated. f ranges from 0 to $4 \times (k - 1)$ and stands for the distance of outermost output bins moving. Although some methods can make f continuous and differentiable [3,9], we still use discrete f. Because f represents a channel choice, and the channel is an integer. Besides this, discrete f can simplify the computation.

For convenient coding and rigorous expression, instead of calculating the location on a feature map after rotation, we calculate which channel is selected at a specific position. And we mark it as $\varphi(i, j \mid f)$. Figure 4(a) is the situation of feature map channel selection for rotational RoI Pooling when offset f equals to 0 and k equals to 5 $(\varphi(i, j \mid 0))$.

9	10	11	12	13
24	1	2	3	14
23	8	0	4	15
22	7	6	5	16
21	20	19	18	17

(a)

15	16	17	18	19
14	4	5	6	20
13	3	0	7	21
12	2	1	8	22
11	10	9	24	23

(b)

Fig. 4. The selected channel of the (i, j)-th bin. The digit of each block is the serial number of a component-sensitive feature map. Different rotations correspond to different selections. Here, we give the examples with offset f equals to 0 (a) and 6 (b) when $k = 5$.

With the definition of $\varphi(i, j \mid 0)$, coupled with the use of mathematical knowledge such as polar coordinates, it is not difficult to derive $\varphi(i, j \mid f)$ which is:

$$\varphi(i, j \mid f) = (\varphi(i, j \mid 0) - (2t - 1)^2 + round(\frac{2tf}{k - 1}))\%n(t) \\ + (2t - 1)^2. \tag{1}$$

in which

$$t = floor((\sqrt{\varphi(i, j \mid 0)} + 1)/2). \tag{2}$$

and

$$n(t) = \begin{cases} 1, & if \quad t = 0, \\ 8t, & otherwise. \end{cases} \tag{3}$$

And a rotational RoI pooling operation in the (i, j)-th bin $(0 \le i, j \le k-1)$ is:

$$r_c(i, j \mid f, \Theta) = \sum_{(x,y) \in bin(i,j)} z_{\varphi(i,j \mid f),c}(x + x_0, y + y_0 \mid \Theta)/n. \tag{4}$$

We follow some definitions in [2] here. $r_c(i, j)$ is the response in the (i, j)-th bin for the c-th category, $z_{\varphi(i,j \mid f),c}$ is one of component feature maps, (x_0, y_0) is the coordinate of left-top corner of an RoI, $\lfloor i\frac{w}{k} \rfloor \le x < \lceil (i + 1)\frac{w}{k} \rceil$, $\lfloor j\frac{h}{k} \rfloor \le y < \lceil (j + 1)\frac{h}{k} \rceil$ (w is the width of an RoI and h is the height of an RoI) and n is the total number of pixels in the (i, j)-th bin.

The $k \times k$ rotation-sensitive scores then give the final response of the RoI. In this paper we simply vote by averaging the scores, producing a $(C + 1)$-dimensional vector for each RoI:

$$r_c(f, \Theta) = \sum_{i,j} r_c(i, j \mid f, \Theta)/(k \times k) \qquad (5)$$

Here, we consider objects possess axial symmetry. Therefore, we do not take mirror operation and the rotation of mirror images into consideration. Note that in this paper, all the rotations are anticlockwise. We give the result of offset 6 when $k = 5$ (Fig. 4(b)).

3.2 Parallel Rotational Feature Extraction Modules

In RR-FCN, we choose the maximum of all rotational RoI pooling output as the final response. And we find that the feature extraction of all rotation angles on the same set of feature maps will cause translation variance fading away. Region proposals in different locations should get different responses. However, since we adopt rotational RoI pooling, the responses may be similar or partly same by rotating. Here are the analysis and solution.

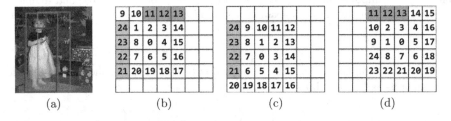

Fig. 5. (a) is the image to be detected with region proposals of red, blue and purple. (b) (c) and (d) are possible responses to the rotational RoI pooling of three region proposals, respectively. Here, $f_b = 0$, $f_c = 15$, $f_d = 2$. We mark out the same response in pure color.

Conflict with Translational Variance. Figure 5 is a simplified region-based detection situation with only three proposals. We can see if proposals are close to each other, they may get partly same outputs. And since we select the maximum response, the probability of this situation is high. So the total number of conflicts is larger even though every two proposals have a little overlap. Therefore, if we train a network with rotation information on the same feature maps, the output bins with different predicted results but same response value will conflict with themselves and lead to the unstable loss oscillation. However, if the difference of two offsets is a multiple of $(k-1)$, they will share no response. That means there are *up to four* kinds of rotational RoI pooling on a blank of component-sensitive feature maps. Thus we use rotational RoI pooling with discrete offset f to ensure that the difference is fixed. By this method, the translation variance can be kept in our network.

Parallel Rotational Feature Extraction Modules. As is analysed above, the available offsets are limited on a blank of component-sensitive feature maps. Thus we make the feature maps and rotational RoI pooling into a feature extraction module (shown in Fig. 2) which can be added in parallel easily. Each module has an initial offset f_i ($0 \leq i < k$). Because the difference of every two offsets should be a multiple of $(k-1)$, the offsets of four rotational RoI pooling layers are f_i, $f_i + (k-1)$, $f_i + 2(k-1)$ and $f_i + 3(k-1)$. Therefore, there are four rotation-sensitive outputs in a module, and we select the maximum of these four as the ouput of this module. The four RoI pooling layers here share feature maps and thus come free of cost which is quite different from maxout [6]. In RR-FCN, we have several feature extraction modules in parallel with different initial f_i to cover more rotational situations and select the max response as the output, too. Therefore, the final rotation-sensitive response in RR-RCN is:

$$r_c(\Theta) = \max_f \; r_c(f, \Theta) \tag{6}$$

Then we use softmax function to normalize the responses across categories.

Interceptive Back Propagation. As we will introduce in Sect. 3.3, we train our networks with only one offset (equals to 0) first and with rotational feature extraction modules in the next step. In order to make our feature maps component-sensitive, we make changes to back propagation in extraction module. We make the network focus on rare situations rather than ones. Table 1 shows the back propagation selections in feature extraction modules. When a region proposal is an object (e.g a dog) and it chooses offset 0 in extraction modules, RR-FCN will ignore this back propagation. In other cases, they run as usual.

Table 1. The back propagation selections in feature extraction module. Note that objects here are class-aware.

	Class-aware object (yes)	Class-aware object (no)
Offset (0)	×	✓
Offset (otherwise)	✓	✓

3.3 Rotational Region-Based Fully Convolutional Networks

Architecture. We take ResNet-101 as our backbone. Since RR-FCN shares the same physical significance with R-FCN [2] when offset f equals to 0, we refer to its structure and make our transformation (we choose the python version code py-R-FCN[1]). As shown in Fig. 2, we use parallel rotational feature extraction modules to get rotation-sensitive responses. However, the bounding box regression method is different. Because bounding boxes are more relative to features without correcting rotation, we use position-sensitive RoI pooling [2] for bounding boxes.

---

[1] https://github.com/YuwenXiong/py-R-FCN.

Training. As is in [5], we define a multi-task loss function on each RoI as $\mathcal{L} = \mathcal{L}_{cls} + \mathcal{L}_{reg}$. $\mathcal{L}_{cls}$ is the cross-entropy loss and $\mathcal{L}_{cls}$ is the smooth L1 loss defined in [5]. Firstly, we set the hyper-parameters the same as [2] and use the model pre-trained on ImageNet [16] to train our network with only a single rotational RoI pooling layer whose offset f equals to 0 for $110k$ iterations. Then we train RR-FCN with rotational feature extraction modules based on the model we get previously. If there are more than one rotational feature extraction module, we make convolution parameters of their component-sensitive feature maps all initialized from the previous model. We use a learning rate of 0.0001, a weight decay of 0.0005 and a momentum of 0.9 to train our model 20k iterations. It is worth mentioning, because we do not know which proposal is rotated and every proposal is important for our network equally, we do not use online hard example mining (OHEM) [17]. Instead of OHEM, we use the interceptive back propagation (Sect. 3.2).

4   Experiments

4.1   Detection Results

We evaluate our methods on PASCAL VOC datasets, and the results are shown in Table 2. We give the detection results of R-FCN [2] model we used in the first step. Besides, we conduct two experiments with different number of rotation offsets (i.e. different number of parallel rotational feature extraction modules). We note the RR-FCN with one rotational feature extraction module as RR-FCN$_4$ (initial offset: 0) and the RR-FCN with two modules as RR-FCN$_8$ (initial offsets: 0 and 1).

As is shown in Table 2, our method achieves state-of-the-art accuracy. And more importantly, feature maps in RR-FCN become component-sensitive. It makes our networks figure out objects' structures and detect successfully under special circumstances (e.g., rotation).

Table 2. Comparisons on PASCAL VOC 2007 and 2012 using ResNet-101. Timing is evaluated on a single NVIDIA TITAN XP, 300 RoIs per image. †: http://host. robots.ox.ac.uk:8080/anonymous/EAVGYV.html ‡: http://host.robots.ox.ac.uk:8080/ anonymous/LU6RUU.html

	mAP(%) (VOC07)	test time (sec/img) (VOC07)	mAP(%) (VOC12)	test time (sec/img) (VOC12)
R-FCN	78.83	0.093	74.50	0.097
RR-FCN$_4$	77.75	0.093	73.31†	0.097
RR-FCN$_8$	77.26	0.102	73.12‡	0.102

4.2   Analysis

Robustness Analysis. Since we construct several blanks of component-sensitive maps in RR-FCN, our model is more robust in real world. Instead of matching directly, RR-FCN can adjust the feature extraction method according to the responses and give more robust predictions.

In order to better illustrate the robustness of the RR-FCN, we detect on the freestyle motorbikes [4] test set[2]. It contains a hundred pictures which contain 128 motorbikes with planar rotations. The test results are shown in Table 3. The RR-FCN$_4$ improves AP by 13.7% which shows that the RR-FCN has better stability under abnormal conditions. Here we give some of detection results (Fig. 6). In these pictures, objects are not in their usual position.

Table 3. Test results on the freestyle test set. All of these networks here are still trained on PASCAL VOC 2007 and 2012.

	R-FCN	RR-FCN$_4$	RR-FCN$_8$
AP(%)	62.4	76.1	74.4

As we can see in Fig. 6, RR-FCN is more accurate under rotation operations. Even these postures are not usual in training set, RR-FCN can still deal it well. That means our network is more robust and it can *comprehend* the object structure.

Feasibility Analysis. RR-FCN learns rotation information from dataset without extra supervision information. We further analyse the convergence of our model. We introduce the ability of RR-FCN to detect objects in rare postures in previous section. The "*rare*" is opposite to "*normal*". For example, we are used to seeing people standing, thus we know what handstand looks like even though we rarely see it. A more mathematical statement is: If I rotate myself x degrees and find a "*normal*" object, in fact, the object is in a "*rare*" posture by rotating itself x degrees. In our experiment, we make our RR-FCN learn what is "*normal*" on tens of thousands of images from PASCAL VOC datasets and then get the rotation information through component-sensitive feature map and rotational RoI pooling. Moreover, we develop interceptive back propagation which can stop our network tolerating rotations.

Necessity Analysis. In the test phase, we may use the previous networks to achieve similar detection results by rotating images. But we argue that it is not appropriate. Firstly, it costs several times of RR-FCN computation. Secondly, because we do not know the rotation angles of the images, the data preprocessing must be unpersuasive. More seriously, there might be wrong results when

---

[2] http://www.iri.upc.edu/people/mvillami/files/iri_freestyle_motocross_dataset_v1.1.
zip#opennewwindow.

(a) failed
motorbike 0.84

(b) motorbike 0.79
motorbike 0.80 & 0.70

(c) person 0.88
person 0.85 & 0.96

(d) aeroplane 0.43
chair 0.85

(e) person 0.98
person 0.85

(f) failed
person 0.9 & bike 0.92

Fig. 6. Some detection results for explaining the robustness of the proposed model. We choose objects in unusual positions and angles to detect. In each group of pictures, the top is the detection result of R-FCN, and the bottom is the detection result of RR-FCN. (a) and (b) are come from freestyle motorbike dataset, (c) and (d) are shot normally, (e) and (f) are come from PASCAL VOC but rotated. We can find RR-FCN shows advantages in detecting objects without the usual position or angle assumption.

detecting rotated images, so we cannot merge the detection results coming from rotated images ideally.

We further analyse the necessity of training. RR-FCN guides the training according to the responses of feature maps. The basic assumptions of our experiments is that the network has been trained well with offset 0 and the most objects appear with usual postures. Under these two assumptions, the trained feature maps in the first step perhaps can detect objects in rare postures with rotational RoI pooling layers. We construct the network with this idea, however, the experiments show that the results are not good. We also test the accuracy of the network (Table 4) and the result is not good, either.

Because the large capacity of deep networks, even the rare objects may have large response in normal pooling method. So we must train the network to make the feature maps component-sensitive.

Table 4. Detection results on PASCAL VOC 2007 using ResNet-101. RR-FCN$_1$ is the network trained only with offset equals to 0, RR-FCN$_{4test}$ is still the network RR-FCN$_1$ but using 4 offsets (0, 6, 12, 18) to test it.

	RR-FCN$_1$	RR-FCN$_{4test}$
mAP(%)	78.83	70.99

5   Conclusion and Future Work

This paper presents a Rotational Region-based Fully Convolutional Network, which is a robust detection network incorporating rotational invariance. We develop a novel pooling method which can share rotational information across different feature maps. To train our networks with the purpose, we propose interceptive back propagation. Moreover, we figure out the specific requests for constructing fully convolution detection networks. In this way, our RR-FCN can learn rotational invariance of objects and detect well even under the situations rarely appear in dataset. That means, It will work better in real world.

We will specify the rotation angle to guide detection and other computer vision task in our future work.

References

1. Boureau, Y., Ponce, J., Lecun, Y.: A theoretical analysis of feature pooling in visual recognition, pp. 111–118 (2010)
2. Dai, J., Li, Y., He, K., Sun, J.: R-fcn: object detection via region-based fully convolutional networks. In: Advances in Neural Information Processing Systems, pp. 379–387 (2016)
3. Dai, J., et al.: Deformable convolutional networks (2017)
4. Fergus, R., Perona, P., Zisserman, A.: Object class recognition by unsupervised scale-invariant learning. In: 2003 Proceedings of the IEEE Computer Society Conference on Computer Vision and Pattern Recognition, vol. 2, pp. II-264–II-271 (2003)
5. Girshick, R.: Fast r-cnn. In: Proceedings of the IEEE International Conference on Computer Vision, pp. 1440–1448 (2015)
6. Goodfellow, I.J., Warde-Farley, D., Mirza, M., Courville, A., Bengio, Y.: Maxout networks. arXiv preprint arXiv:1302.4389 (2013)
7. He, K., Zhang, X., Ren, S., Sun, J.: Spatial pyramid pooling in deep convolutional networks for visual recognition. In: Fleet, D., Pajdla, T., Schiele, B., Tuytelaars, T. (eds.) ECCV 2014. LNCS, vol. 8691, pp. 346–361. Springer, Cham (2014). https://doi.org/10.1007/978-3-319-10578-9_23

8. He, K., Zhang, X., Ren, S., Sun, J.: Deep residual learning for image recognition. In: Proceedings of the IEEE Conference on Computer Vision and Pattern Recognition, pp. 770–778 (2016)
9. Jaderberg, M., Simonyan, K., Zisserman, A., Kavukcuoglu, K.: Spatial transformer networks. In: Neural Information Processing Systems, pp. 2017–2025 (2015)
10. Lin, T.Y., Dollár, P., Girshick, R., He, K., Hariharan, B., Belongie, S.: Feature pyramid networks for object detection. arXiv preprint arXiv:1612.03144 (2016)
11. Liu, W., et al.: SSD: single shot multibox detector. In: Leibe, B., Matas, J., Sebe, N., Welling, M. (eds.) ECCV 2016. LNCS, vol. 9905, pp. 21–37. Springer, Cham (2016). https://doi.org/10.1007/978-3-319-46448-0_2
12. Long, J., Shelhamer, E., Darrell, T.: Fully convolutional networks for semantic segmentation. In: Computer Vision and Pattern Recognition, pp. 3431–3440 (2015)
13. Redmon, J., Divvala, S., Girshick, R., Farhadi, A.: You only look once: Unified, real-time object detection. In: Proceedings of the IEEE Conference on Computer Vision and Pattern Recognition, pp. 779–788 (2016)
14. Redmon, J., Farhadi, A.: Yolo9000: better, faster, stronger. arXiv preprint arXiv:1612.08242 (2016)
15. Ren, S., He, K., Girshick, R., Sun, J.: Faster r-cnn: towards real-time object detection with region proposal networks. In: Advances in Neural Information Processing Systems, pp. 91–99 (2015)
16. Russakovsky, O., et al.: Imagenet large scale visual recognition challenge. Int. J. Comput. Vis. **115**(3), 211–252 (2015)
17. Shrivastava, A., Gupta, A., Girshick, R.: Training region-based object detectors with online hard example mining. In: IEEE Conference on Computer Vision and Pattern Recognition, pp. 761–769 (2016)

Face Detection for Crowd Analysis Using Deep Convolutional Neural Networks

Bryan Kneis[✉]

University of West England, Bristol, BS16 1QY, UK
bryan2.kneis@live.uwe.ac.uk

Abstract. Crowd analysis is a challenging topic within computer vision, current state of the art methods for face detection in crowds suffer from poor results due to visual occlusions, scene semantics and overlapping subjects. In this work, we propose a novel approach of utilizing existing semantic segmentation methods to detect and segment faces in obscured images. We use an implementation of Mask RCNN trained on the popular Labelled Faces in the Wild (LFW) database to compare performance with Viola Jones, histogram of orientated gradients and max-margin object detection using a synthetically generated occluded subset of LFW. Results show that when images contain fair sized occlusions, Mask RCNN outperforms the current state of the art method. State of the art performance was achieved on this dataset and context specific improvements are suggested for further work. The contribution of this paper is not to regurgitate the finding from the original paper on Mask RCNN but provide results on the efficiency of using the method in the context of face detection for crowd analysis. Additionally, exploration of suitable hyper parameters for this context has been performed and described. Code has been made publicly available.

Keywords: Deep learning · Face detection · Crowd analysis

1 Introduction

As our population grows cities become denser and larger amounts of congregations of people occur on a daily basis. Humans are social creatures and when in close proximity highly dynamic crowd behaviour occurs [15]. Swarm intelligence is an example of the types of intelligent behaviours that can emerge when biological entities coordinate in close spaces [16]. Current methods for crowd analyses are relatively novel and provide poor results [17], and as such, applications within the domain are limited. Currently most crowd analysis applications use equipment such as people counters which track people from a bird's eye view. This spatial information can then be aggregated to log the movement of people [1]. Unfortunately, this type of setup is not applicable for a vast number of crowd analysis applications, mainly in the surveillance and security sectors.

People counters only identify the position of a subject in a crowd. This type of data is extremely useful retrospectively, when analysed later for crowd management, but poor for responsive tasks, such as surveillance. By detecting faces in a crowd more identifiable information can be sourced to be used in a responsive manner, for instance,

© Springer Nature Switzerland AG 2018
E. Pimenidis and C. Jayne (Eds.): EANN 2018, CCIS 893, pp. 71–80, 2018.
https://doi.org/10.1007/978-3-319-98204-5_6

tracking down a person of interest in a scene. Additionally, people trackers are limited as they cannot be deployed in constrained environments such as underground stations, due to low ceilings.

As the world population grows, more occurrences of phenomena relating to crowd behaviour emerges [18] and as such changes need to be made to the study and preventative measures of our crowds. By including biometric information, such as a face images, we can cater for types of applications not possible with just spatial information. It is this combination of biometric and spatial information that can capture the dynamic social behaviour of a crowd in addition to the information useful to act upon it.

The method detailed in this work attempts to increase the robustness of face detection of occluded images usually found in crowd analysis applications. In addition to this the use of segmenting the face could be useful for three other contexts. Firstly, segmenting the face during detection can simplify the pre-processing required for face recognition. Second, the data captured from the segmented images could be aggregated for the use of super resolution techniques and face hallucination [2]. Lastly, using the segmented image could improve the robustness of blob matching algorithms used in face tracking. By matching only the segmented faces the amount of noise would be reduced and correspondence between frames is more likely.

2   Related Work

2.1   Face Detection in Unconstrained Environments

Sliding window based methods have been widely adopted for the task of object detection [19–22]. They allow emergent features to be uncovered as the scale of data being analysed changes. The Viola Jones algorithm uses this approach for face detection, where a sliding window is iterated over an image to detect haar features. Since the development of Viola Jones several variants have been proposed for improved performance in unconstrained environments [28]. Most of these methods utilize some handcrafted features for the classifier, in contrast to the standard haar features.

Zhang et al. proposed a multi resolution framework for general object detection [24]. During detection the framework uses lower resolution features to reject the majority of negative windows. This allows for more resource to be used to process the remaining windows in higher resolution. Combing this approach with standard histogram of gradient features [25] produced better results than the traditional sliding windows approach.

Focused on achieving state of the art performance without the requirement of large amounts of training data, Trivedi et al. propose a boosted decision tree classifier for face detection [3]. Using a standard soft cascade, the method attempts to greedily minimize the loss function of the classifier. Results were attained on the FDDB dataset [14], although performance was comparable to some deep neural networks, the boosted decision trees saturated early when provided more data. Unlike CNN's that consistently benefit from additional modelling capacity [3], the classifier was unable to gain further performance improvements. This lead to a peak performance below other state of the art methods.

2.2 Convolutional Neural Network Based Face Detection

In 1994 Vaillant et al. [5] applied convolutional neural networks to the task of face detection. The approach consisted of two steps, first the neural network was presented with a pixel and its neighbourhood. The network then performs a rough localization to determine if the area indicated the presence of an object. In the second step, these areas are presented to another neural network to determine the exact position of the object.

Li et al. was the first to achieve state of the art performance on the face detection data base (FDDB) [14] in 2015 using CNN's [4]. The method uses a cascade approach where the image is analysed at different resolutions. The network first rejects areas considered background at the lower resolution, similar to the multi resolution framework in [24]. After this a calibration stage occurs in which the window positions of potential objects are adjusted for the subsequent detection stage.

Fast RCNN [6] coined the term region based convolutional neural network. Although the methods leading up to Fast RCNN used region proposal methods, Fast RCNN was the first to identify it as the distinguishing factor in its performance. Using the deep VGG16 network the method achieved state of the art performance on PASCAL VOC 2012 [7] data set challenge.

Advances from Fast RCNN [6] and SPP net [8] in detection network runtime lead to the region proposal algorithm becoming a bottle neck [9]. To overcome this problem Faster RCNN [9] was developed and introduced the Region Proposal Network (RPN). This approach shared the convolutional layers of the RPN with the detection network, thus enabling almost cost-free region proposals.

Region proposal algorithms are an approach to sub sample an image to avoid computation of the entire image. Max-margin object detection (MMOD) [27] takes a different approach in which it optimizes all the windows for its detection. Results 4 show it is beating the state of the art method Faster RCNN [29] using the FDDB database.

In this paper we intend to evaluate the performance of both traditional feature-based detection and CNN based methods including Mask RCNN [10] on our synthetically occluded database. These results will then aid in further discussion on the limitations of the methods and where they can be improved for the context of crowd analysis.

3   Implementation

Our method uses an implementation of Mask RCNN [30] to generate the bounding boxes and instance masks of a face. Mask RCNN is a generalized approach capable of supporting different backbone architectures for feature extraction of the image and the network head for mask prediction. We chose to use the popular resnet-101 architecture [26] for the former and feature pyramids network for the latter. This decision was predominantly made using the results of the original paper in [10]. Furthermore, independent research on different backbone networks in [11] show resnet-101 outperforming competing architectures such as VGG in several different tasks. Training of the model was performed on the well-known Labelled Faces in the Wild (LFW) dataset [12] using the subset containing annotated ground truth labels. From our research this seemed the only usable data set for training a deep CNN as it had over 2900 annotated images.

Training of the network was performed in two stages, first the head layers are trained. Freezing the resnet backbones layers the region proposal network, classifier and mask heads are trained. Next the rest of the layers are added, and the network is trained end to end using a learning rate of 0.001 for 2 epochs with 1000 steps per epoch.

To reduce memory load the masks are downscaled to 65×65. Due to inaccuracies of labelling ground truth data at scale the effect on performance is negligible. Figure 1 shows the original image and mask on the left, as well as the downscaled image and mask in the middle. The re projected mask and image is shown on the right, where pixel loss can be observed around the edges of the mask.

Fig. 1. Process of downscaling instance masks

Region Proposal Networks (RPN) were first introduced in [10] with the development of Faster RCNN. An RPN is a fully convolutional network that simultaneously predicts object bounds and confidence scores at set anchor positions. The anchors define the scales and aspect ratios of the data being analysed. We opted to use a set of anchors for the RPN at 8, 16, 32, 64 and 128. The stride of the anchors differs from the original paper [10] set at 4px. We chose to use a stride of 2 as this showed not to compromise accuracy while enjoying a reduced computation and memory load. By doing this the number of anchors used was cut by 4, resulting in 16368 anchors total per image (Fig. 2).

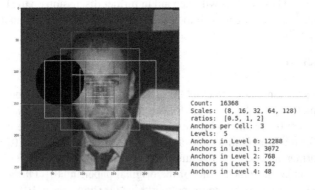

```
Count:  16368
Scales:  (8, 16, 32, 64, 128)
ratios:  [0.5, 1, 2]
Anchors per Cell:  3
Levels:  5
Anchors in Level 0: 12288
Anchors in Level 1: 3072
Anchors in Level 2: 768
Anchors in Level 3: 192
Anchors in Level 4: 48
```

Fig. 2. RPN Anchors on centre pixel of image (left), statistics of anchors (right)

In order to extract the best anchors, we ranked them based on their proposal of the object and took the best ones. These best anchors are called region of interests (ROI), for our network we have chosen to use 32 ROI's to output from the RPN. We chose this number since we have only 2 classes to predict, face and background, and because the images were small. Shown in Fig. 3 are the ROI's outputted from the RPN projected onto the image. These 32 images are then fed into the object classifier, using the image from Fig. 3 we find that 10 ROI's are positive and 22 are negative.

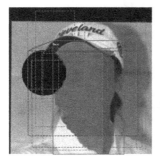

```
Positive ROIs:  10
Negative ROIs:  22
Positive Ratio: 0.31
Unique ROIs: 32 out of 32
```

Fig. 3. Visual of positive ROI's for image (left), statics of ROI (right)

Occlusions in the dataset have been generated synthetically; this allows variation in the amount of occlusion to determine the relationship to the methods performance. A circle was generated at a random position in the image to occlude the face. Typically, occlusions in the context of crowd analysis are continual, so our choice of occlusion was preferred over other methods such as salt and pepper noise. Furthermore, the position of the occlusion was randomly determined, as shown in Fig. 4, to achieve an average performance over a range of occlusions.

Fig. 4. Randomly occluded images

To validate the performance of Mask RCNN the occlusion placed on the image needs to be present in the object mask. This ensures that the occlusion does not contribute to the area of the intersection over union. Figure 5 shows images with an occlusion and their corresponding masks.

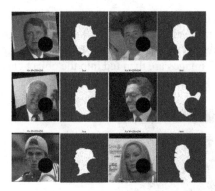

Fig. 5. Synthetic occlusions generated on images and ground truth labels

4   Results

4.1   Face Detection on Occluded Images

We perform a comparison between Mask RCNN, Viola Jones, HOG features combined with a linear classifier and the max-margin object detection algorithm implemented using a CNN on our synthetically occluded subset of the Labelled Faces in the Wild. Provided is the AP (Average Precision) at a threshold of 0.5 across a range of occlusion amounts. The performance metric has been chosen based on the evaluation criteria of other object detection datasets, such as PASCAL VOC 2012 [7]. The amount of occlusion has been varied and plotted to indicate the relationship between the occlusion of a face and detection performance. The amount of occlusion was varied by the radius of the circle, starting from 0 and ending at 90px.

All our results were produced on the same computer using an nvidia gtx 970 m and Intel i7 CPU. When evaluating a training validation split of 80/20 was used, training the model on 2600 images it took around 4 h.

Mask RCNN outperforms all other methods on every validation set, as shown in Fig. 6. Although performance is comparable when the occlusion amount is very low, performance degradation can be seen much quicker within the alternative methods as the occlusion amount increases. These results support our hypothesis that Mask RCNN out performs the current state of the art methods for occluded images.

4.2   Face Recognition on Segmented Images

Shown in Fig. 7 are the results of the face recognition performance when performed on the original LFW database. Figure 8 plots the results of the same methods using the segmented LFW database where each image was segmented using Mask RCNN. Results were obtained using an off the shelf open source face recognition system, OpenFace [13]. OpenFace was chosen as it has shown state of the art performance on the LFW dataset and provides sufficient documentation for reproducing results. Our results showed that the recognition rate was comparable with a marginally small drop using the

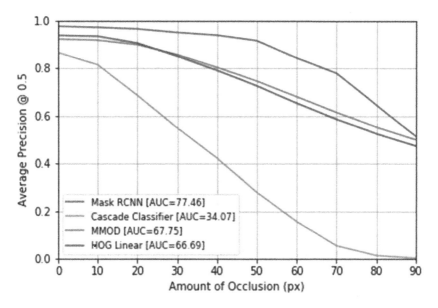

Fig. 6. Face detection performance using AP@0.5 for varying occlusion amounts

segmented images. This not only proves that our segmentation method does not drastically affect recognition performance but illustrates how well the CNN's are able to determine the useful features of a face and inherently ignore background.

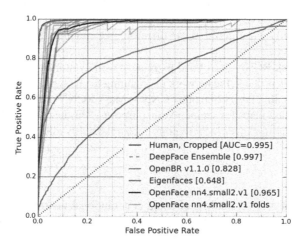

Fig. 7. Open face performance on LFW

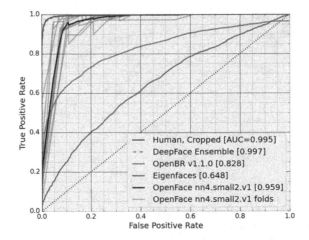

Fig. 8. Open face performance on segmented LFW

5   Discussion

As shown in the results section Mask RCNN outperforms all of the other methods evaluated. Of all the methods Viola Jones suffered the worse results when provided occlusions. We believe the reason for this is the necessity of cascading features. When some of the features are missing due to the occlusion, the cascade will cause the classifier to fail. The linear classifier using HOG features and MMOD performed relatively well under occlusion. Intuitively the performance of any algorithm will diminish when missing semantically meaningful data, therefore we assume these two algorithms performed as expected.

In contrast Mask RCNN performed extremely well even under severe occlusions. We hypothesize this is due to the network maintaining the feature's spatial information. Instead of using pooling layers like other CNN's, Mask RCNN preserves the alignment of the image and class maps for segmentation purposes. This leads to a holistic analysis of the data were missing small local regions does not harshly effect performance.

6   Conclusion

We have presented a novel method of using convolutional neural networks typically used for semantic segmentation and applied them to the problem domain of obscured face detection. Advances in convolutional neural networks for object detection like 10 Fast RCNN and R-FCN have improved bounding box regression and detection accuracies. It was not until Mask RCNN that allowed parallel computation of object masks to overcome the shortcomings in design so that it could be used for semantic segmentation. Results show that our method outperforms the current state of the art in obscured images, but when occlusions are not present or very little, the algorithms performance are comparable.

In further work we intend on implementing a face tracking system where the segmentation method is used to replace the current face detection method. At present face tracking suffers from poor results when faces are obscured, especially when the obscurity is of another face. If the faces are segmented from the occlusion, correspondence between frames could be improved and therefore increase the robustness of the face tracking algorithm.

References

1. Leonardi, F., Marcii, D.: An uncertainty model for people counters based on video sensors. In: Advanced Methods for Uncertainty Estimation in Measurement, Italy, pp. 62–66 (2008)
2. Liu, C., Shum, H.Y., Freeman, W.: Face hallucination: theory and practice. Int. J. Comput. Vision **75**(1), 115–134 (2007)
3. Eshed, O.B., Trivedi, M.: To boost or not to boost? On the limits of boosted trees for object detection. In: 23rd International Conference on Pattern Recognition (ICPR), pp. 3350–3355, Mexico (2016)
4. Li, H., Lin, Z., Shen, X., Brandt, J., Hua, G.: A convolutional neural network cascade for face detection. In: Proceedings of the IEEE Conference on Computer Vision and Pattern Recognition, Boston, pp. 5325–5334 (2015)
5. Vaillant, R., Monrocq, C., Le Cun, Y.: Original approach for the localisation of objects in images. In: IEE Proceedings-Vision, Image and Signal Processing, pp. 245–250 (1994)
6. Girshick, R.: Fast R-CNN. arXiv preprint (2015). arXiv:1504.08083
7. Everingham, M., Van Gool, L., Williams, C., Winn, J., Zisserman, A.: The PASCAL Visual Object Classes Challenge 2012 (VOC2012) Results. http://host.robots.ox.ac.uk/pascal/VOC/voc2012
8. He, K., Zhang, X., Ren, S., Sun, J.: Spatial pyramid pooling in deep convolutional networks for visual recognition. In: Fleet, D., Pajdla, T., Schiele, B., Tuytelaars, T. (eds.) ECCV 2014. LNCS, vol. 8691, pp. 346–361. Springer, Cham (2014). https://doi.org/10.1007/978-3-319-10578-9_23
9. Ren, S., He, K., Girshick, R., Sun, J.: Faster R-CNN: towards real-time object detection with region proposal networks. In: Advances in Neural Information Processing Systems, pp. 91–99 (2015)
10. He, K., Gkioxari, G., Dollár, P., Girshick, R.: Mask R-CNN. In: IEEE International Conference on Computer Vision (ICCV), Venice, pp. 2980–2988 (2017)
11. Lin, T.Y., Dollár, P., Girshick, R., He, K., Hariharan, B., Belongie, S.: Feature pyramid networks for object detection. In: Conference on Computer Vision and Pattern Recognition (CVPR). vol. 1, no. 2, pp. 4–13, Hawaii (2017)
12. Gary, B., Marwan, M., Honglak, L.: Erik Learned-Miller: Labeled Faces in the Wild: A Database for Studying Face Recognition in Unconstrained Environments. http://vis-www.cs.umass.edu/lfw/#reference
13. Amos, B., Bartosz, L., Satyanarayanan, M.: OpenFace: A general-purpose face recognition library with mobile applications. https://cmusatyalab.github.io/openface
14. Vidit, J., Erik, LM.: FDDB: A Benchmark for Face Detection in Unconstrained Settings. http://vis-www.cs.umass.edu/fddb
15. Kinjal, J., Safvan, V.: Crowd behavior analysis. Int. J. Sci. Res. (IJSR), **3**(12) (2014)
16. Garnier, S., Gautrais, J., Theraulaz, G.: The biological principles of swarm intelligence. Swarm Intell. **1**(1), 3–31 (2007)

17. Junior, J.C.S.J., Musse, S.R., Jung, C.R.: Crowd analysis using computer vision techniques. IEEE Signal Process. Mag. **27**(5), 66–77 (2010)
18. Moussaïd, M., Perozo, N., Garnier, S., Helbing, D., Theraulaz, G.: The walking behaviour of pedestrian social groups and its impact on crowd dynamics. PloS one. **5**(4) (2010)
19. Girshick, R., Donahue, J., Darrell, T., Malik, J.: Rich feature hierarchies for accurate object detection and semantic segmentation. In: Proceedings of the IEEE Conference on Computer Vision and Pattern Recognition, Ohio, pp. 580–587 (2014)
20. Sermanet, P., Eigen, D., Zhang, X., Mathieu, M., Fergus, R., LeCun, Y.: Overfeat: Integrated recognition, localization and detection using convolutional networks. arXiv preprint (2013). arXiv:1312.6229
21. Wang, X., Han, T.X., Yan, S.: An HOG-LBP human detector with partial occlusion handling. In: IEEE 12th International Conference on Computer Vision, pp. 32–29, Kyoto (2009)
22. Maji, S., Malik, J.: Object detection using a max-margin hough transform. In: IEEE Conference on Computer Vision and Pattern Recognition (CVPR), pp. 1038–1045, Miami (2009)
23. Redmon, J., Divvala, S., Girshick, R., Farhadi, A.: You only look once: unified, real-time object detection. In: Proceedings of the IEEE conference on computer vision and pattern recognition, pp. 779–788, Las Vegas (2016)
24. Zhang, W., Zelinsky, G., Samaras, D.: Real-time accurate object detection using multiple resolutions. In: 11th International Conference on Computer Vision (ICCV), pp. 1–8 (2007)
25. Dalal, N., Triggs, B.: Histograms of oriented gradients for human detection. In: Computer Vision and Pattern Recognition (CVPR), vol. 1, pp. 886–893 (2005)
26. He, K., Zhang, X., Ren, S., Sun, J.: Deep residual learning for image recognition. In: Proceedings of the IEEE Conference on Computer Vision and Pattern Recognition, pp. 770–778, Las Vegas (2016)
27. King, D.E.: Max-margin object detection. arXiv preprint (2015) arXiv:1502.00046
28. Barr, J.R., Bowyer, K.W., Flynn, P.J.: The effectiveness of face detection algorithms in unconstrained crowd scenes. In: IEEE Winter Conference on Applications of Computer Vision (WACV), pp. 1020–1027, Hayden (2014)
29. King D.: Easily Create High Quality Object Detectors with Deep Learning. http://blog.dlib.net/2016/10/easily-create-high-quality-object.html
30. Waleed, K.: Mask R-CNN for object detection and instance segmentation on Keras and TensorFlow. https://github.com/matterport/Mask_RCNN

Extreme Learning Machine and
Machine Learning Applications

Smoothing Regularized Extreme Learning Machine

Qin-Wei Fan[1(✉)], Xing-Shi He[1], and Xin-She Yang[2]

[1] School of Science, Xi'an Polytechnic University,
Xi'an 710048, People's Republic of China
`qinweifan@126.com`
[2] School of Science and Technology, Middlesex University London,
The Burroughs, London NW4 4BT, UK

Abstract. Extreme learning machines have been applied successfully to many real-world applications, due to their faster training speed and good performance. However, in order to guarantee the convergence of the ELM algorithm, it initially requires a large number of hidden nodes. In addition, extreme learning machines have two drawbacks: over-fitting and the sensitivity of accuracy to the number of hidden nodes. The aim of this paper is to propose a new smoothing $L_{1/2}$ extreme learning machine with regularization to overcome these two drawbacks. The main advantage of the proposed approach is to reduce weights to smaller values during the training, and such nodes with sufficiently small weights can eventually be removed after training so as to obtain a suitable network size. Numerical experiments have been carried out for approximation problems and multi-class classification problems, and preliminary results have shown that the proposed approach works well.

Keywords: Neural networks · Extreme learning machine (ELM)
Smoothing $L_{1/2}$ regularization · Sparsity

1 Introduction

Feedforward neural networks (FNNs) are a class of the most popular neural networks with a diverse range of applications in many fields [1–5]. Such neural networks have very good capabilities for approximation and multiclass classification [6–11]. But the training of such neural networks lacks faster learning algorithms, and traditional learning algorithms are usually far slower than required, which is a major bottleneck for neural networks and their applications [12,13].

The extreme learning machine (ELM), proposed in [12,14], is a new learning algorithm for single hidden layer feedforward networks (SLFNs), and this algorithm partly overcomes the problems of learning algorithms for FNNs concerning slow convergence and multimodality. In contrast with the most practical

Q.-W. Fan—This work was supported by National Science Foundation of China (No. 11171367).

E. Pimenidis and C. Jayne (Eds.): EANN 2018, CCIS 893, pp. 83–93, 2018.
https://doi.org/10.1007/978-3-319-98204-5_7

implementations that all the parameters of the FNNs need to be tuned, the ELM does not necessarily adjust the input weights and its first hidden layer biases in applications, thus ELM can be a much faster learning machine with better generalization performance than other learning algorithms under certain conditions [11–14,16]. However, the stochastic nature of the hidden layer output matrix may potentially lower the learning accuracy of ELMs. In addition, it has been observed that, because of the random selection of input weights and hidden node biases, a large number of hidden nodes may be required to achieve a certain level of accuracy.

In order to overcome these drawbacks, regularization methods have been proposed in the literature. For example, a L_1-norm based approach was proposed [16,17], so as to make some of the fitted coefficients of the model become exactly zero, hence leading to sparse models that are easily interpretable. To overcome the issue of over-fitting, an L_2 regularized ELM was proposed by penalizing the training errors [18,19]. But in order to retain a satisfactory accuracy, the L_2 regularized ELM usually requires a much higher number of hidden neurons, which results in a complicated network [15,18,20].

More recently, the sparsity problem has attracted much attention, and researchers aim to find sparse solutions of a representation or an equation. Given an $M \times N$ matrix A, the sparsity problem can often be transformed into the following L_0 regularization problem:

$$\min\ \{\|y - Ax\|^2 + \lambda\|x\|_0\}, \tag{1}$$

where $x = (x_1, x_2, \cdots, x_N)^T \in \mathbb{R}^N$, and λ is a penalty or regularization parameter. Here, $\|x\|_0$ is called an L_0-norm which is the number of non-zero components of x. The main purpose of this model is to recover or invert x from a set of observations y such that x has the fewest nonzero components. Though an L_0 regularizer can yield the most sparse solution, it is essentially a combinatorial optimization problem, which can be very challenging to solve.

In order to overcome this issue, an L_1 regularization based approach was proposed [21–23], which tries to minimize

$$\min\ \{\|y - Ax\|^2 + \lambda\|x\|_1\}, \tag{2}$$

where $\|x\|_1$ is the L_1-norm of $\mathbb{R}^N$. However, for many practical applications, the solutions of the L_1 regularizer are often less sparse than those of the L_0 regularizer [28]. Thus, researchers have developed the following, general L_r ($0 < r < 1$) regularization [24–26], which minimizes

$$\min\ \{\|y - Ax\|^2 + \lambda\|x\|_r^r\}, \quad 0 < r < 1, \tag{3}$$

where $\|\cdot\|$ denotes the Euclidean norm of $x = (x_1, x_2, \cdots, x_N)^T \in \mathbb{R}^N$. The L_r-norm is defined by

$$\|x\|_r = \Big(\sum_{i=1}^{N} |x_i|^r\Big)^{1/r}. \tag{4}$$

Now the main question is that what value of r should be used so as to obtain the best result for difficult types of problems? In [27–30], an $L_{1/2}$ regularizer has been proposed and shown to have a few promising properties such as unbiasedness, sparsity and oracle properties. They also showed that the $L_{1/2}$ regularizer could be easier to solve than the L_0 regularizer, with a higher sparsity than a L_1 regularizer. Thus, $L_{1/2}$ can be taken as a representative of the L_r $(0 < r < 1)$ regularizer, which has the following objective:

$$\min\ \{\|y - Ax\|^2 + \lambda \|x\|_{1/2}^{1/2}\}. \tag{5}$$

It is worth pointing out that, though $L_{1/2}$ regularizer have some desirable properties, the $L_{1/2}$ regularization term involves absolute values and is not differentiable at the origin, which typically causes oscillations during the training. To address these issues, a smoothing regularized algorithm has been proposed [31–33], and they showed that their modified algorithms can dampen oscillations and also have an effect on sparsity. However, there are still many issues that need to be addressed further.

In this paper, a smoothing $L_{1/2}$ regularizer for ELM is proposed. The aim of this paper is to enhance the performance of the ELM by introducing a new smoothing $L_{1/2}$ regularization term so as to prevent the weights from taking too large values. One of the effects of this additional smoothing term is to intentionally force certain unnecessary weights to zero during the training process so as to reduce the size of the large neural network to a suitable size of the trained neural networks. Therefore, the rest of this paper is organized as follows. Section 2 introduces briefly the fundamental of an extreme learning machine. Section 3 presents the $L_{1/2}$ regularization of ELM, and Sect. 4 discusses two case studies for evaluating the performance of the proposed approach. Finally, Sect. 5 concludes with some discussions.

2   Fundamentals of an Extreme Learning Machine

An extreme learning machine (ELM) works for generalized single hidden layer feedforward networks (SLFNs), which was proposed by Huang et al. [12–15]. The essence of an ELM is that the weights for the hidden layer nodes of SLFNs need not to be tuned and the output weights can be calculated by using the least-squares method. Their studies show that ELM provides better generalization performance at a much faster learning speed than those traditional learning techniques.

For a set of N arbitrary distinct samples $(x_i, t_i) \in \mathbb{R}^N \times \mathbb{R}^M$, the basic model concerns the relationship between inputs $x_i = (x_{i1}, x_{i2}, \cdots, x_{in}) \in \mathbb{R}^N$ and the corresponding desired output values o_i. The ELM model approximation first starts with the randomly assigned values for the weight vector $a_i = (a_{i1}, a_{i2}, \cdots, a_{in})^T \in \mathbb{R}^N$, connecting the ith hidden node and the input nodes. The activation is done in terms of $b_i \in \mathbb{R}$ as the threshold of the ith hidden node. Let β_i be the weight vector connecting the ith hidden node with the output nodes. Then, an ELM with L hidden nodes and activation function $g(x)$ can be expressed in the following form:

$$f_L(x_j) = \sum_{i=1}^{L} \beta_i \, g(a_i \cdot x_j + b_i) = o_j, \quad (j = 1, 2, \cdots, N), \tag{6}$$

which can be written more compactly in a matrix form

$$\mathbf{H}\,\beta = \mathbf{O}. \tag{7}$$

Here, the matrices are given by

$$\mathbf{H}(a_1, \cdots, a_L, b_1, \cdots, b_L, x_1, \cdots, x_N) = \begin{bmatrix} g(a_1, b_1, x_1) & \cdots & g(a_L, b_L, x_1) \\ \vdots & \ddots & \vdots \\ g(a_1, b_1, x_N) & \cdots & g(a_L, b_L, x_N) \end{bmatrix}_{N \times L} \tag{8}$$

$$\beta = \begin{bmatrix} \beta_1^T \\ \vdots \\ \beta_L^T \end{bmatrix}_{L \times M}, \quad O = \begin{bmatrix} o_1^T \\ \vdots \\ o_N^T \end{bmatrix}_{N \times M}, \quad T = \begin{bmatrix} t_1^T \\ \vdots \\ t_N^T \end{bmatrix}_{N \times M}. \tag{9}$$

This model can be formulated as an optimization problem by minimizing

$$||\beta||_r + \lambda||\mathbf{H}\beta - \mathbf{T}||_s, \tag{10}$$

where $r, s > 0$ and λ is a penalty or regularization parameter. This form is what most variants of ELMs have used. The ELM approach is to initialize randomly the values of a_i and b_i, and inverse parameter estimation is usually done by using the Moore-Penrose pseudo-inverse [34] so as to estimate output weights $\beta = H^\dagger O$.

3 Smoothing the ELM with an $L_{1/2}$ Regularizer

In order to optimize the input weight matrix and hidden layer weights, as described in the introduction part, the $L_{1/2}$ regularizer has many desirable properties such as unbiasedness, sparsity and oracle properties, especially for its sparse effect on the network structure. In the ELM framework, we will replace the conventional error function by a smoother regularization term using an $L_{1/2}$-norm, which leads to a smoothing ELM regularizer (SELMR). The introduction of this $L_{1/2}$ term is to smooth the ELM such that it prevents the weights from taking too large values, and thus forcing unnecessary weights to approach to zero during the training. Consequently, their corresponding hidden neurons can be removed so as to reduce the network to a suitable size.

The error function with a smoothing $L_{1/2}$ regularizer becomes the following form:

$$E(\beta) = \frac{1}{2} \sum_{k=1}^{N} ||o_k - t_k||^2 + \sum_{i=1}^{L} \sum_{j=1}^{M} f(\beta_{ij})^{\frac{1}{2}}, \tag{11}$$

where $f(x)$ is a smooth function that approximates $|x|$. For removes the oscillation of the gradient value and get better pruning results, we can choose $f(x)$ as a piecewise polynomial function:

$$f(x) = \begin{cases} |x|, & \text{if } |x| \geq a, \\ -\frac{x^4}{8a^3} + \frac{3x^2}{4a} + \frac{3a}{8}, & \text{if } |x| < a. \end{cases} \tag{12}$$

It is straightforward to show that

$$f(x) \in [\frac{3}{8}a, +\infty), \tag{13}$$

and its derivatives are in the range of

$$f'(x) \in [-1, 1], \quad f''(x) \in [0, \frac{3}{2a}], \tag{14}$$

where a is a very small positive constant (close to zero).

In addition, it is easy to verify that the gradient of the error function with respect to β_{ij} is given by

$$\frac{\partial E(\beta)}{\partial \beta_{ij}} = \sum_{k=1}^{N} g(a_i, b_i, x_k)(o_{kj} - t_{kj}) + \lambda \frac{f'(\beta_{ij})}{2f(\beta_{ij})^{1/2}}. \tag{15}$$

and λ is a penalty or regularization parameter. By using the notation for the increment

$$\Delta\beta_{ij} = -\frac{\partial E(\beta)}{\partial \beta_{ij}}, \quad (i = 1, 2, \cdots, L; j = 1, 2, \cdots, M), \tag{16}$$

the SELMR algorithm has the following iterative updating equation for weights β^n

$$\beta_{ij}^{n+1} = \beta_{ij}^n + \eta\Delta\beta_{ij}, \tag{17}$$

and the iterations typically starts with an arbitrary, initial weight vector β^0. Here, $n(= 0, 1, 2, ...)$ is the iteration counter and η is the learning rate.

4   Performance Evaluation

In order to evaluate the performance of the proposed method in this section, two different sets of simulations are presented, and the comparison of compare the proposed SELMR with the standard ELM algorithm has been carried out using a number of available benchmark datasets for regression, and multi-class classification in the literature. We have run the used algorithms on the same data sets, and each algorithm has been executed for 30 runs with the same fixed size for all algorithms. The average results of 30 trial simulations are obtained and presented in the rest of this paper.

4.1   Example 1: Function Regression Problem

To evaluate the effectiveness of the proposed SELMR algorithm in this section, we consider the following commonly used function-fitting benchmark:

$$F(x) = 1.1(1 - x + 2x^2)e^{-x^2/2}, \quad x \in [-4, 4]. \tag{18}$$

The shape of this function is shown in Fig. 1. Comparison of the performance of the approximation capability of the proposed SELMR with that of the standard ELM algorithm will also be carried out.

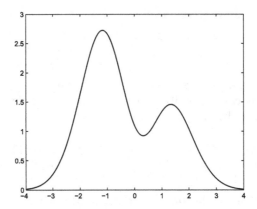

Fig. 1. Benchmark function as the approximation target.

In this numerical experiment, a set of 1000 input sampling points x_i, ($i = 1, 2, ..., 1000$) are randomly selected from a uniform distribution in the interval $(-4, 4)$. Then, 500 input points are stochastically selected from the 1000 points as the training set, while the remaining 500 points are used for validation and testing generalisation. To make the fitting problem harder, all the training data points are perturbed with a level of noise that is uniformly distributed in $[-0.1, 0.1]$.

Both algorithms have used 60 nodes whose weights are randomly initialized from a uniform distribution in $[-1, 1]$. The learning rate is $\eta = 0.09$ and regularization parameter is $\lambda = 0.005$. Each algorithm has been run for 30 times, and the results have been averaged and then analyzed statistically.

Figures 2 and 3 show the results and comparison. From both figures, we can clearly see that the SELMR algorithm can approximate the benchmark function very well. In addition, the results in terms of RMSE values are summarized in Table 1 where we can see that SELMR algorithm is better than ELM algorithm. More importantly, the average number of hidden nodes of SELMR is much fewer than that of ELM. This means that SELMR gives better pruning; namely, the final weights to be removed are smaller than those produced through the usual $L_{1/2}$ regularization.

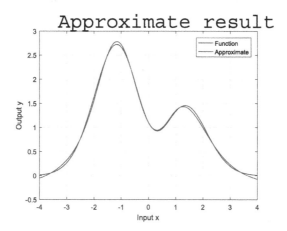

Fig. 2. Approximation by ELM.

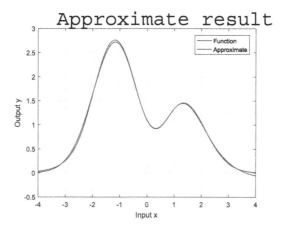

Fig. 3. Approximation by SELMR.

Table 1. Performance comparison of ELM and SELMR for Example 1.

Learning algorithms	Hidden nodes	Training RMSE	Testing RMSE
ELM	60	0.0727	0.0521
SELM	34.16	0.0702	0.0439

This multimodal test problem is still relatively simple. In order to validate the proposed algorithm further, let us use a multi-class classification problem as a second case study.

4.2   Example 2: Benchmark Classification Problem

There are many benchmarks for classifications, and we have selected 6 classification data sets from the standard UCI machine learning data repository[1], and they are summarized in Table 2. These data sets have then been tested by ELM and SELMR, respectively. In this example, each dataset is randomly split into two parts: 60% used for testing, while the remaining 40% as the training set.

Table 2. Benchmark classification data sets for Example 2.

Data sets	Data size	Input features	Classes
Iris	150	4	3
Wine	178	13	3
Diabetes	768	8	2
Ecoli	336	7	8
Splice-junction	3190	61	3
Landsat	6435	36	3

To compare the computational performance, the original learning rate for both algorithms has been set to be 0.2. For the SELMR, the penalty parameter has been set to be 0.003 for each classification data set. As before, 30 independent runs have been carried out for each data set. The average results are summarized in Table 3.

Table 3. Comparison between ELM and SELMR for Example 2.

Data set	ELM			SELMR		
	Training accuracy	Test accuracy	Nodes	Training accuracy	Test accuracy	Nodes
Iris	0.9381	0.8937	50	0.9541	0.9230	30.36
Wine	0.9832	0.9772	100	0.9912	0.9826	60.64
Diabetes	0.8530	0.8329	200	0.8523	0.8340	126.82
Ecoli	0.7863	0.7681	150	0.8091	0.7820	79.30
Splice-junction	0.8910	0.8760	1000	0.8912	0.8711	593.60
Landsat	0.9063	0.8781	1000	0.9182	0.8920	601.30

Three performance metrics have been listed in Table 3 where we can clearly see that SELMR can achieve a higher precision performance with fewer nodes for almost all cases, except for the diabetes data where both methods obtained

---

[1] https://archive.ics.uci.edu/ml/datasets.html.

comparable results. Even so, the number of nodes used by SELMR is about 1/3 fewer than that used by ELM. All these results indicate that the proposed approach can indeed reduce the trained networks to a suitable size and yet be able to approximate and classify the datasets with a better accuracy.

5   Conclusion

Though ELM based approaches can be effective, they do have some drawbacks. In this paper, we have introduced a smoothing $L_{1/2}$ regularization term to the standard ELM algorithm so as to overcome existing drawbacks. The proposed SELMR has been tested using a function approximation benchmark and 6 multi-class classification datasets. The simulation results show that our proposed SELMR can optimize the network structure to a suitable size with fewer hidden nodes and also obtain better accuracy than the standard ELM.

Though the preliminary results are promising, further studies are needed. For example, a more detailed parametric study is needed to figure out the optimal ranges for hyper-parameters such as the learning rate, the regularization parameter and the number of initial hidden nodes. In addition, more extensive validation with other data sets is also useful. Furthermore, theoretical analysis of the proposed SELMR in terms of convergence properties can gain better insight into the algorithms.

Acknowledgement. This work has been supported by National Science Foundation of China (No. 11171367).

References

1. Haykin, S.: Neural Networks: A Comprehensive Foundation, 2nd edn. Tsinghua University Press, Prentice Hall, Beijing (2001)
2. Magoulas, G.D., Vrahatis, M.N., Androulakis, G.S.: Improving the convergence of the backpropagation algorithm using learning rate adaptation methods. Neural Comput. **11**(7), 1769–1796 (1999)
3. Zhang, X.S.: Neural Networks in Optimization. Kluwer Academic Publishers, Boston (2000)
4. Liu, W., Dai, Y.H.: Minimization algorithms based on supervisor and searcher cooperation. J. Optim. Theory Appl. **111**(2), 359–379 (2001)
5. Zhou, W., Zurada, J.M.: Competitive layer model of discrete-time recurrent neural networks with LT neurons. Neural Comput. **22**(8), 2137–2160 (2010)
6. Poggio, T., Girosi, F.: A theory of networks for approximation and learning. Artificial Intelligence Laboratory, Mass. Inst. Technol., Cambridge, A.I. Memo 1140 (1989)
7. Hornik, K., Stinchcombe, M., White, H.: Multilayer feedforward networks are universal approximators. Neural Netw. **2**, 359–366 (1989)
8. Cybenko, G.: Approximation of superpositions of a sigmoidal function. Math. Contr. Signals Syst. **4**(2), 303–314 (1989)
9. Funahashi, K.: On the approximate realization of continuous mappings by neural networks. Neural Netw. **2**, 183–192 (1989)

10. Hornik, K.: Approximation capabilities of multilayer feedforward networks. Neural Netw. **4**, 251–257 (1991)
11. Huang, G.B., Chen, Y.Q., Babri, H.A.: Classification ability of single hidden layer feedforward neural networks. IEEE Trans. Neural Netw. **11**(3), 799–801 (2000)
12. Huang, G.B., Zhu, Q.Y., Siew, C.K.: Extreme learning machine: theory and applications. Neurocomputing **70**, 489–501 (2006)
13. Huang, G.B., Wang, D.H., Lan, Y.: Extreme learning machines: a survey. Int. J. Mach. Learn. Cyber. **2**, 107–122 (2011)
14. Huang, G.B., Zhu, Q.Y., Siew, C.K.: Extreme learning machine: a new learning scheme of feedforward neural networks. In: Proceedings of the IEEE International Joint Conference on Neural Networks, vol. 2, pp. 985–990 (2004)
15. Huang, G.B., Zhou, H.M., Zhang, R.: Extreme learning machine for regression and multiclass classification. IEEE Trans. Syst. Man Cybern. Part B Cybern. **42**(2), 513–529 (2012)
16. Balasundaram, S., Kapil, D.G.: 1-Norm extreme learning machine for regression and multiclass classification using Newton method. Neurocomputing **128**, 4–14 (2014)
17. Zhang, L., Zhou, W.: On the sparseness of 1-norm support vector machines. Neural Netw. **23**, 373–385 (2010)
18. Deng, W., Zheng, Q., Chen, L.: Regularized extreme learning machine. In: Proceedings of the IEEE Symposium on Computational Intelligence in Data Mining, pp. 389–395 (2009)
19. Miche, Y., van Heeswijk, M., Bas, P., Simula, O., Lendasse, A.: TROP-ELM: a double-regularized ELM using LARS and Tikhonov regularization. Neurocomputing **74**, 2413–2421 (2011)
20. Luo, J.H., Vong, C.M., Wong, P.K.: Sparse Bayesian extreme learning machine for multi-classification. IEEE Trans. Neural Netw. Learn. Syst. **25**(4), 836–843 (2014)
21. Candes, E., Romberg, J., Tao, T.: Stable signal recovery from incomplete and inaccurate measurements. Commun. Pure Appl. Math. **59**(8), 1207–1223 (2006)
22. Donoho, D.L.: Compressed sensing. IEEE Trans. Inf. Theory **52**(4), 1289–1306 (2006)
23. Donoho, D.L.: Neighborly polytopes and the sparse solution of underdetermined systems of linear equations. Statistics Department, Stanford University, Stanford, CA, Technical report 2005-4 (2005)
24. Chartrand, R., Staneva, V.: Restricted isometry properties and nonconvex compressive sensing. Inverse Probl. **24**(3), 20–35 (2008)
25. Krishnan, D., Fergus, R.: Fast image deconvolution using hyper-Laplacian priors. In: Neural Information Processing Systems. MIT Press, Cambridge (2009)
26. Chartrand, R.: Exact reconstruction of sparse signals via nonconvex minimizaion. IEEE Signal Process. Lett. **14**(10), 707–710 (2007)
27. Xu, Z.B., Zhang, H., Wang, Y., et al.: L1/2 regularization. Sci. China Inf. Sci. **53**, 1159–1169 (2010)
28. Xu, Z.B., Chang, X.Y., Xu, F.M., Zhang, H.: $L_{1/2}$ regularization: a thresholding representation theory and a fast solver. IEEE Trans. Neural Netw. Learn. Syst. **23**(7), 1013–1027 (2012)
29. Zeng, J.S., Fang, J., Xu, Z.B.: Sparse SAR imaging on $L_{1/2}$ regularization. Sci. China Inf. Sci. **55**(8), 1755–1775 (2012)
30. Meng, D.Y., Zhao, Q., Xu, Z.B.: Improve robustness of sparse PCA by L_1-norm maximization. Pattern Recognit. **45**(1), 487–497 (2012)

31. Fan, Q.W., Zurada, J.M., Wu, W.: Convergence of online gradient method for feedforward neural networks with smoothing $L_{1/2}$ regularization penalty. Neurocomputing **131**, 208–216 (2014)
32. Wu, W., Fan, Q.W., Zurada, J.M., et al.: Batch gradient method with smoothing L1/2 regularization for training of feedforward neural networks. Neural Netw. **50**, 72–78 (2014)
33. Yang, D.K., Wu, W.: A Smoothing Interval Neural Network, Discrete Dynamics in Nature and Society, vol. 2012, 25p (2012)
34. Rao, C.R., Mitra, S.K.: Generalized Inverse of Matrices and Its Applications. Wiley, New York (1971)

Neuroevolution of Actively Controlled Virtual Characters - An Experiment for an Eight-Legged Character

Svein Inge Albrigtsen, Alexander Imenes, Morten Goodwin,
Lei Jiao[✉], and Vimala Nunavath

Centre for Artificial Intelligence Research, University of Agder,
4879 Grimstad, Norway
thhethssmuz@gmail.com, alexander.imenes@gmail.com,
{morten.goodwin,lei.jiao}@uia.no

Abstract. Physics-based character animation offers an attractive alternative for traditional animations. However, it is often strenuous for a physics-based approach to incorporate active user control of different characters. In this paper, a neuroevolutionary approach is proposed using HyperNEAT to combine individually trained neural controllers to form a control strategy for a simulated eight-legged character, which is a previously untested character morphology for this algorithm. It is aimed to evaluate the robustness and responsiveness of the control strategy that changes the controllers based on simulated user inputs. The experiment result shows that HyperNEAT is able to evolve long walking controllers for this character. In addition, it also suggests a requirement for further refinement when operated in tandem.

Keywords: Artificial intelligence · Neuroevolution
Actively controlled virtual characters · Eight-legged character

1 Introduction

Character animation has become an important part of modern game development. The responsiveness and perceived realism of these animations play a key role for providing an immersive experience to the players. While an experienced animator is often able to create lifelike and realistic looking animations, the limitation of these animations is also obvious as they are only applicable for the purpose that the animation portrays. When the animations are utilized outside of their intended domain, however slightly, they start to fall short.

Physics-based simulation offers an attractive alternative to traditional animation techniques, wherein each motion is the direct result of a physics simulation and is therefore physically realistic by definition [5]. Physics-based animations are commonly used to simulate passive phenomena like objects, cloths, fluids and ragdolls. However, for more active animations most games still resort to kinematics-based approaches [5,7,11]. One of the commonly cited reasons is

© Springer Nature Switzerland AG 2018
E. Pimenidis and C. Jayne (Eds.): EANN 2018, CCIS 893, pp. 94–105, 2018.
https://doi.org/10.1007/978-3-319-98204-5_8

that physics-based simulated characters are notoriously difficult to control, as all movement has to be controlled by the application of torques and/or other forces [5]. One way to tackle this issue is to train controllers using machine learning, a technique that has shown promising results [1,2,6,8–10,12–15]. However, such controllers are usually trained for one singular purpose, or an action, and has little application outside of that purpose. While one approach could be to train multiple controllers and then switch between them, the actual implementations of this logic is sparse. It is therefore not known whether these controllers could handle the incessant switching that would be required inside a highly interactive environment like a game. In this paper, we investigate the feasibility of combining individually trained neural controllers to form a control strategy for actively controlled virtual characters. The proposed approach taken to this end utilizes neuroevolution to train a small set of neural controllers. These controllers will be utilized to generate joint torques for a physics-based simulated multi-legged character to produce motion for a corresponding set of targeted behaviors. In addition, the efficiency of these controllers can be evaluated based on how robust they perform when switching among them. The main goal of this study is twofold. Firstly, we evaluate whether HyperNEAT is able to evolve gaits for an eight-legged character, which, to the best of our knowledge, has not been studied before. Secondly, we study whether neural controllers trained for different behaviors can operate in tandem to produce robust interactive controllers.

The contributions of this study mainly reside in the approach taken to train these networks, by using HyperNEAT to generate gaits for a previously untested, highly complex, eight-legged character. We also provide insight into the efficiency of constructing high-level controllers using such evolved networks. The unabridged version[1] of this work, the source code[2] of this study and the videos[3] are all available online.

The remainder of this paper is organized as follows. In Sect. 2, the approach of the experiment is detailed. The numerical results are given in Sect. 3 before we conclude the work in the last section.

2 The HyperNEAT Based Approach

This section details the approach and methods applied in this study. More specifically, the design of the character model and substrate, and most notably the setup of the physics simulation are presented.

2.1 Multi-legged Character

Before we explain the multi-legged character, the environment for our study is introduced presently. In this study, we use a custom OpenGL-based game engine that has been adapted to incorporate a physics engine and neuroevolution library

---

[1] http://hdl.handle.net/11250/2454827.

[2] https://github.com/reewr/master.

[3] https://reewr.github.io/master.

in order to simulate and train the controllers. The engine itself is mainly used for visualization and to control the flow of the physics-based simulation. The Bullet3 physics engine[4] is adopted to perform all rigid body simulation and collision detection required to simulate the character. To evolve the controllers, the neuroevolution library MultiNEAT[5] is utilized. The choice of MultiNEAT in particular is made due to language compatibility with the rest of the code.

Fig. 1. The arachnoid used in this study.

In the aforementioned game engine, an arachnoid with a robotic theme, featuring eight legs each of which is equipped with five different joints, is utilized in our study, as shown in Fig. 1. This model is chosen as it has a sufficient complexity to provide a realistic example of a fully featured game character. All the joints of the character are defined as hinge joints, meaning that they can only rotate in one specific axis. Furthermore, all these joints have limitations in how much they can move in their respective axises. This is to enforce realistic motions as the joints should not have a full 360-degree freedom of movement. Figure 2 shows each joint and their limits. As seen in Fig. 2, the Trochanter is the only joint that is able to rotate forwards or backwards in relation to the sternum, whereas every other joint can only rotate up and down. The actual values of these limits can be seen in Table 1. The character has four additional parts that are not mentioned in Table 1, the head, neck, hip and abdomen, which may also be rotated. However, these parts are not considered to contribute much in terms of the movement of the character and therefore have been disabled and set to a constant angle that should not interfere with the movement of the character.

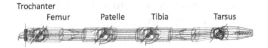

Fig. 2. Leg with joint limits.

The dimensions of the character are roughly 4 units from its head to the back of the abdomen and has approximately 9 units in total leg span when fully

---

[4] www.bulletphysics.org.
[5] www.multineat.com.

stretched out. Each part of the character has a weight of 1 mass unit, except for the sternum, abdomen and head which weighs 10 mass units, 8 mass units and 5 mass units respectively. These units are configured to make sure that the center of mass is not exactly at the center of the character, but rather slightly tilted towards the back, in order to increase the perceived realism of the character.

Table 1. Limits of the joints of the character's legs in degrees.

Joint	Upper limit	Lower limit
Trochanter	−60	60
Femur	−20	60
Patella	−100	5
Tibia	−100	5
Tarsus	−35	35

2.2 Substrate Applied in HyperNEAT

As mentioned above, HyperNEAT uses a substrate to define the coordinates for input, hidden and output nodes. Designing the substrate is an important aspect of using HyperNEAT as it contributes to how it will determine symmetries and patterns [4].

The substrate is designed to match the geometry of the character as closely as possible to make it easier for HyperNEAT to detect the symmetries of the legs. The substrate takes inspiration from previous research that has evolved gaits for Quadruped [2,3,8,15], but is extended to support the additional two legs on each side and the increased number of joints. The substrate with all its layers can be seen in Fig. 3, which is defined with three two-dimensional 7×8 Cartesian grids. All unmarked inputs/outputs are un-utilized.

Each column in the substrate represents a leg, starting from the front left and going clockwise around the character. Each row, except for the two top-most rows, represents the current angle of a joint normalized from a value within $[-\pi, \pi]$ to a value within $[-1, 1]$. The second row represents whether the tip of the leg is touching the floor whereas the first row includes the pitch, roll and yaw of the sternum. The first row also includes a sine and cosine wave to encourage periodic behavior. All the inputs, hidden and output coordinates of the substrate are spread uniformly in the range of $[-1, 1]$, while trying to keep symmetry between opposite legs. To differentiate the layers, the inputs, hidden and output layers are placed on different z-coordinates, where the input z-coordinate is 1, hidden z-coordinate is 0 and output z-coordinate is 1.

The outputs are expected to be within the range $[1, 1]$. The current angle of the joint is subtracted from the output and the result of this is set as the new velocity of the joint. This simulates setting a target angle of the joint, allowing the networks to choose their desired angles of each joint.

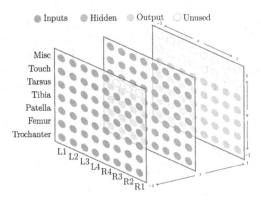

Fig. 3. The three-dimensional substrate of the character.

2.3   Controllers

In order to create the control strategy, the neural controllers for each action has to be trained first.

The first controller is trained to make the character stand completely still in a standing position. While training, it will be awarded for not moving away from the initial starting position and for remaining still in a balanced pose. If the character falls at any point during the simulation, i.e. touches the ground with any vital parts, the simulation will be terminated, and the character will be rewarded for the length of time it stayed alive. This is mainly to discourage such behavior by limiting the fitness of individuals that are unable to remain upright throughout the entirety of the simulation, promoting more stable postures.

The second controller is trained to make the character walk with a stable gait. Working under the assumption that the further the character walks the more stable the gait must be. Therefore, the fitness of the character will be the furthest distance traveled in one specific axis. In this case, it is the positive z-axis, which is the axis that the character is facing at the start of the simulation. As with the first controller, if the character falls, the simulation will stop and the final fitness value will be the furthest distance traveled before it falls.

The reason for choosing these two actions is that together they form a minimum of actions required to test the control strategy. Walking gaits have been evolved for various legged creatures before and thus evolving this behavior for the character used in this project should hopefully not be much of a stretch, considering the characters heightened complexity. While it would easily be possible not to have a separate standing controller, e.g., by just locking all the joints of the character instead, this could possibly cause the character to fall over if it happens to stop in some unbalanced positions. Thus, it is required for a separate standing controller that could account for such imbalance and other residual forces from the walking controller after a transition.

Both controllers are trained via HyperNEAT using the substrate described in the previous section, and use the same input and output scheme as described above.

The simulation process will be the same for both controllers and they will only differ in the fitness functions that they use. All simulations will be run with 150 individuals with randomized weights based on a random seed for 300 generations. Each character will be allowed to run for up to 10 s in real time, in addition to an un-simulated process of one second where the character is positioned into a standing pose that is equal across all simulations.

Due to the complexity of HyperNEAT and its underlying NEAT algorithm, it has a variety of different parameters to set, most of which have been heavily based on previous research for similar multi-legged characters [3,4,8,15]. However, these may be subject to a certain degree of trial-and-error depending on the results. A full list of the NEAT/HyperNEAT parameters may be found in the unabridged version of this work.

2.4 Control Strategy

Once these two controllers are in place, a control strategy can be evaluated by using the above controllers together. The control strategy is designed to measure the responsiveness of the controllers, which is done by using the controllers in sequence and measuring the time that it takes to transition from standing to walking and vice versa.

3 Performance Evaluations

In this section, we will evaluate our proposed approach by two examples, i.e., standing controller and walking controller independently, and then will test the controlling strategy, i.e., by testing them alternating in a tandem manner.

3.1 Standing Controller

The standing controller is evaluated based on its ability to stand still and not falling to the ground.

In early generations of the simulation the dominating factor by far is the killing of individuals who fell. In the first generations rarely does even a single individual live long enough as to complete the full simulation duration. This leads to a dramatic increase in simulation speed in the early stages as it allows the simulation to be cut short and the next generation could be started.

Seeing as the controller is directly rewarded for how long it remains alive, it is not surprising that most of the controllers managed to learn that staying upright is a good strategy. However, many controllers go beyond that and therefore often flail about frantically in order to remain upright. An example of this behavior can be seen in Fig. 4.

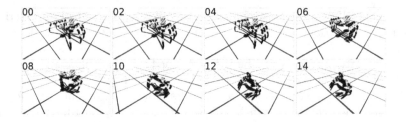

Fig. 4. Image series of the first standing controller.

This controller tries to remain upright by pulling all its legs towards its body, however, it appears to do so with an equal force across all legs. Seeing as the characters center of mass is slightly towards the back, this symmetrical application of force leads the character to tilt, eventually falling backwards.

Other controllers manage to learn that certain poses are easier to balance than others. As such, many controllers evolved various static poses that they are able to hold near indefinitely. However, some of these poses look more like a spider mannequin that has been randomly assembled by a tornado. One example of this can be seen in Fig. 5. This controller does take some time to gather itself, but eventually converges to a stable pose. While certainly amusing, it does not meet the requirements of the control strategy, as it is considered unlikely that a walking controller could naturally transition into this particular pose.

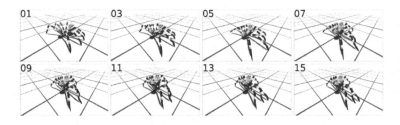

Fig. 5. Image series of another standing controller.

Among many controllers for standing, a controller is selected as the champion, and an image series of the champion controller is shown in Fig. 6. The final controller does not achieve a perfectly balanced pose as it slowly descended towards the ground. However, it never does any drastic motions to position itself and are therefore judged to be the most compatible with other controllers in the control strategy.

Figure 7 illustrates the generated neural network for the champion standing controller. As can be observed, the number of connections is low and they are mostly grouped around the top of the substrate. Since it was rewarded for not moving and only to keep itself from falling, it only had to use some of the joints to do exactly that. Seeing as the number of connections in a network will increase its sensitivity, this sensitivity would usually cause it to move around more. In

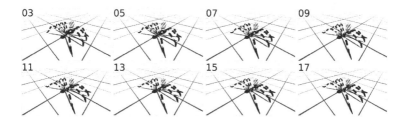

Fig. 6. Image series of the champion standing controller.

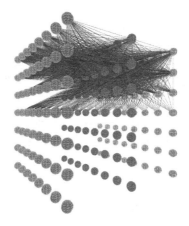

Fig. 7. The generated neural network for the champion standing controller.

this case, HyperNEAT discovers that reducing the number of connections in the network would make it better at standing still. An additional note is that most of the outputs are also not used. The downside of this type of network is that it is very hard for it to react to changes, unless those changes happen to hit the very few input nodes that its connected to.

3.2 Walking Controller

The walking controller was evaluated purely based on how far it traveled in one specific axis while remaining upright. While all the evolved controllers are able to move forward to some degree, classifying all of them as gaits would be a stretch.

The controllers often quickly learned that standing as far up as it could and then falling forward would give it a fairly decent fitness score. In some training runs it learned that by throwing itself forward from the standing position where it started in, it could easily cover a distance of 4–5 units. If evolution discovered any of these behaviors early in the simulation, it would often outcompete other individuals and eventually dominate the population. Further evolution of these controllers would usually only get incrementally better at falling over or throwing themselves forward. An example from such a controller can be seen in Fig. 8.

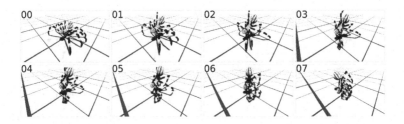

Fig. 8. Image series highlighting the gait of the first walker.

After many trial of different controllers, the champion controller is presented below. This controller, which is not punished for touching the ground, displays the most purposeful and straight movement of all the controllers. The gait exhibits a strong asymmetric tendency in two distinct groups, consisting of the four front and four back legs, resembling a form of gallop. This controller is also capable of sustained walking far beyond the training period, in addition to being stable. The image series of the champion walker is shown in Fig. 9 and the final generated neural network for the champion is illustrated in Fig. 10.

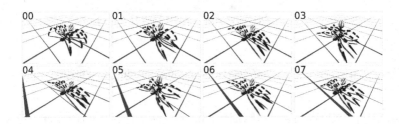

Fig. 9. Image series of the gait of the champion walker.

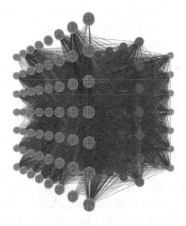

Fig. 10. The generated neural network for the champion walker.

3.3   Controller Strategy

The proposed control strategy is evaluated based on its responsiveness and robustness when switching between the two controllers. For this experiment, the walking and standing controllers are sequenced in a loop with 5 s time intervals, and the velocity of the character is recorded at each update. Figure 11 shows the result of this experiment.

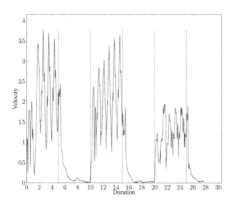

Fig. 11. Velocity of the character over time in seconds. The transitions between walking and standing are indicated by the red vertical lines. (Color figure online)

Measuring the actual responsiveness of this experiment was more difficult than anticipated, as the standing controller never stopped completely. That is to say, its velocity never reached exactly 0 as can be seen in Fig. 11. Since the controller never reaches the threshold value of 0, by using a larger threshold value, it becomes possible to measure the responsiveness when switching from walking to standing. Depending on how strict this threshold value is configured, the average response time is measured to 0.41 s for a lenient threshold velocity of 1 and 1.54 s when the threshold is set to a strict 0.1.

The same measurement is applied for the switching from the standing to the walking controller. The walking controller does not have a constant speed, but rather displays a periodically varying phase of movement and rest. However, measuring the responsiveness with a threshold value is similarly effective, even when the threshold is set higher than the low-points of the resting period of its gait. The average response time for this switch is measured at 0.09 s for a lenient threshold velocity of 0.5 and 0.65 s for a stricter threshold value of 1.5.

As can be seen in Fig. 11, the speed of the walking controller reduces noticeably after the second switch from the standing controller. The reason for this is that it locks one of its forelegs under two of its hind legs.

4   Conclusions

We investigate the feasibility of combining individually trained neural controllers to form a control strategy that could be used to actively control virtual physics-based characters. The HyperNEAT is adopted to evolve neural controllers for a previously untested domain of eight-legged locomotion. To create and evaluate the control strategy, two target neural controllers are trained for standing and walking. The newly trained neural controllers are combined to form the control strategy and is evaluated for its robustness and responsiveness when switching between them. The results show that HyperNEAT is able to evolve gaits for a highly complex eight-legged character. The resulting gaits showed that they are quite capable of walking long distances, even beyond the training period. However, when combined within the control strategy, the results suggest a need for further refinement as the controllers are still not robust enough to operate in tandem.

References

1. Allen, B.F., Faloutsos, P.: Evolved controllers for simulated locomotion. In: Egges, A., Geraerts, R., Overmars, M. (eds.) MIG 2009. LNCS, vol. 5884, pp. 219–230. Springer, Heidelberg (2009). https://doi.org/10.1007/978-3-642-10347-6_20
2. Clune, J., Beckmann, B.E., Ofria, C., Pennock, R.T.: Evolving coordinated quadruped gaits with the hyperneat generative encoding. In: 2009 IEEE Congress on Evolutionary Computation, pp. 2764–2771, May 2009. https://doi.org/10.1109/CEC.2009.4983289
3. Clune, J., Stanley, K.O., Pennock, R.T., Ofria, C.: On the performance of indirect encoding across the continuum of regularity. IEEE Trans. Evol. Comput. **15**(3), 346–367 (2011). https://doi.org/10.1109/TEVC.2010.2104157
4. Clune, J., Ofria, C., Pennock, R.T.: The sensitivity of HyperNEAT to different geometric representations of a problem. In: Proceedings of the 11th Annual Conference on Genetic and Evolutionary Computation, GECCO 2009, pp. 675–682. ACM, New York (2009). https://doi.org/10.1145/1569901.1569995
5. Geijtenbeek, T., Pronost, N.: Interactive character animation using simulated physics: a state-of-the-art review. Comput. Graph. Forum **31**(8), 2492–2515 (2012). https://doi.org/10.1111/j.1467-8659.2012.03189.x
6. Grzeszczuk, R., Terzopoulos, D.: Automated learning of muscle-actuated locomotion through control abstraction. In: Proceedings of the 22nd Annual Conference on Computer Graphics and Interactive Techniques, SIGGRAPH 1995, pp. 63–70. ACM, New York (1995). https://doi.org/10.1145/218380.218411
7. Hagenaars, M.: Hierarchical development of physics-based animation controllers. Master's thesis, Utrecht University (2014)
8. Lee, S., Yosinski, J., Glette, K., Lipson, H., Clune, J.: Evolving Gaits for physical robots with the HyperNEAT generative encoding: the benefits of simulation. In: Esparcia-Alcázar, A.I. (ed.) EvoApplications 2013. LNCS, vol. 7835, pp. 540–549. Springer, Heidelberg (2013). https://doi.org/10.1007/978-3-642-37192-9_54
9. Morse, G., Risi, S., Snyder, C.R., Stanley, K.O.: Single-unit pattern generators for quadruped locomotion. In: Proceedings of the 15th Annual Conference on Genetic and Evolutionary Computation, GECCO 2013, pp. 719–726. ACM, New York (2013). https://doi.org/10.1145/2463372.2463461

10. Olson, R.S.: A step toward evolving biped walking behavior through indirect encoding. Honors in the major thesis, University of Central Florida (2010)
11. Pejsa, T., Pandzic, I.: State of the art in example-based motion synthesis for virtual characters in interactive applications. Comput. Graph. Forum **29**(1), 202–226 (2010). https://doi.org/10.1111/j.1467-8659.2009.01591.x
12. Reil, T., Husbands, P.: Evolution of central pattern generators for bipedal walking in a real-time physics environment. IEEE Trans. Evol. Comput. **6**(2), 159–168 (2002). https://doi.org/10.1109/4235.996015
13. Sims, K.: Evolving virtual creatures. In: Proceedings of the 21st Annual Conference on Computer Graphics and Interactive Techniques, SIGGRAPH 1994, pp. 15–22. ACM, New York (1994). https://doi.org/10.1145/192161.192167
14. Valsalam, V.K., Miikkulainen, R.: Modular neuroevolution for multilegged locomotion. In: Proceedings of the Genetic and Evolutionary Computation Conference, GECCO 2008, pp. 265–272. ACM, New York (2008). http://nn.cs.utexas.edu/?valsalam:gecco08
15. Yosinski, J., Clune, J., Hidalgo, D., Nguyen, S., Zagal, J.C., Lipson, H.: Evolving robot gaits in hardware: the HyperNEAT generative encoding vs. parameter optimization. In: Proceedings of the 20th European Conference on Artificial Life, pp. 890–897 (2011)

Machine Learning with the Pong Game: A Case Study

Benedikt Nork[1], Geraldine Denise Lengert[1], Robert Uwe Litschel[1], Nasim Ahmad[1], Gia Thuan Lam[2], and Doina Logofătu[1(✉)]

[1] Department of Computer Science and Engineering,
Frankfurt University of Applied Sciences, 60318 Frankfurt a.M., Germany
`logofatu@fb2.fra-uas.de`
[2] Vietnamese German University, Le Lai Street, Thu Dau Mot City, Vietnam

Abstract. Being one of the earliest computer games, the Pong game is well-known for its simplicity, which makes it suitable for becoming one of the very first problems in Artificial Intelligence and Machine Learning: The goal is to create a self-playing agent that can compete against humans. In the past there have been introduced various Machine Learning approaches to solve this problem. This paper gives a summary of some notable techniques to creating a self-learning agent for the Pong game. In addition, it proposes a template for developing this idea into a full-fledged application. An implementation in Java is available online.

Keywords: Pong game · Machine Learning · Self-playing agent

1 Introduction

Invented by Atari [1], Pong is a 2-dimensional computer game inspired from the well-known sport table tennis [5,6]. Pong features a simple graphical user interface, with a circle representing a ball and two rectangle bricks on two opposite sides representing the players [2,8]. The shape of the bricks and the ball may vary in accordance with a designer's taste, but one property that remains unchanged over the course of time is simplicity: Each player can invoke only 2 actions, either to move up or to move down. Thanks to its simplicity, Pong, as well as other Atari games, has become a test subject since the early days of Machine Learning and Artificial Intelligence [4,11], fields specializing in creating self-functioning computer programs that can act depending on the environment, and does things which used to be doable only by humans such as playing computer games. In this study, our focus is only on the Pong game. Our paper gives and discusses some notable Machine Learning algorithms that have been applied for this problem. Furthermore, we will provide a template for developing it into a real computer program with real implementation available on our website [7]. This can serve as a starting point for researchers with interest in the field of Machine Learning.

© Springer Nature Switzerland AG 2018
E. Pimenidis and C. Jayne (Eds.): EANN 2018, CCIS 893, pp. 106–117, 2018.
https://doi.org/10.1007/978-3-319-98204-5_9

2   Previous Work

In the following we present an overview of two notable Machine Learning solutions to the Pong problem.

2.1   Neural Network with Backpropagation

In general the backpropagation algorithm consists of the following steps:

– **Calculating the forward phase** – calculating the output of the neural network.
– **Calculating the backward phase** – calculating the error term for each layer in the network starting from the last and using the results to backpropagate to the first one.
– **Combining the individual gradients** – yields the total gradient for all input-output neuron pairs.
– **Updating the weights** – using the learning rate α and the previously determined total gradient.

An adequate initial learning rate α has to be chosen and the weight between each pair of neurons can be randomly initialized before the start of the algorithm.

The backpropagation network used in our work is based on that of Tariq Rashid [9]. The network starts with randomly selected edge weights between -1 and 1. The function needed to update the edge weights is:

$$- (t_k - o_k) \cdot sigmoid \left(\sum_j w_{jk} \cdot o_j \right) \left(1 - sigmoid \left(\sum_j w_{jk} \cdot o_j \right) \right) \cdot o_j \quad (1)$$

The first part $-(t_k - o_k)$ is the error term. t_k is the value to be achieved and o_k is the last calculated value. The sum within the sigmoid functions is the signal in the nodes of the last layer. It is just the signal in a node before the activation function is applied. The last part o_j is the output from node j of the previous hidden layer. In this process you begin in the output layer and end in the input layer.

2.2   Neural Network with Evolutionary Algorithm

Typically, even in a simple neural network one would find a considerable amount of parameters that can be tuned to enhance the network's overall performance. Often this is done using brute force techniques or the developer's previous experience and better judgement. However, it is possible to apply a heuristic approach, such as an evolutionary algorithm, that can offer significant improvements regarding effort costs, but obtains similar results as the brute force. In its essence an evolutionary approach resembles evolution in the real world:

– A set of randomly generated solutions is created, which in the context of evolution represents a population of individual.

- Each solution is evaluated to determine its adequateness, i.e. the fitness of the individual.
- The best solutions (fittest individuals) are selected to generate a new and hopefully better set (the next generation).
- New solutions are generated by combining old ones (much like parents producing offspring) or by altering old ones (much like mutation as we know it).
- The same process is repeated with the new set of solutions (generation) and gradually the overall fitness of the candidate solutions should increase. After a fixed number of iterations or after a given fitness value is reached, the best solution is chosen and it represents an optimal approximation of the best solution to the problem in question.

In the context of neural networks parameters adequate for tuning include, but are not limited to: (1) the number of layers; (2) the number of neurons per layer; (3) the dense layer activation function; (4) the network optimizer and others. The general approach involves several steps:

- Initialize a population of N randomly generated networks.
- Evaluate each network by training it and analyzing its performance in solving the task in question.
- Sort the networks according to their fitness and use those with the highest score to generate the next generation. In this step different techniques may be employed, e.g. elitism (bringing several of the very best individuals from the parent generation into the new generation).

When creating the new generation of networks the parameters of the new individuals can be mutated, effectively resulting in better or worse tuning. This approach would eventually generate a population, where the fittest individual contains an approximation of the best values for the network parameters, which we can now use for our actual neural network.

An example implementation has been introduced by Rinaldi's [10]. It starts with randomly selected edge weights between −1 and 1 and a predetermined number of genomes per generation. Each time the ball hits the brick, a score counter is increased by one. Every time the brick misses the ball a new genome is tested. When all genomes have been tested, a new generation is created with the following possibilities:

- If none of the genomes scored a point, it will once again randomize a whole new generation of genomes.
- If a genome has scored a point, it will create a new generation, but it will mix the new generation with the genome that has scored the point.
- If several genomes have scored a point, it will create a new generation, but the new generation will be mixed with the best genome of the old generation (Fig. 1).

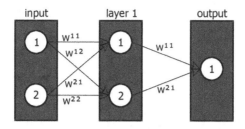

Fig. 1. Structure of the two networks. Two input neurons in the input layer (neuron 1 gets the y-value of the bricks, neuron 2 gets the y-value of the ball), two neurons in the first and only hidden layer and an output neurone. Each neuron of a layer is connected to every neuron in the next layer.

3   Implementation Details

In the following subsections, we will give a description of all the element of our application which is available online [7].

3.1   Description

In this project, a program has been created in Java. It allows human playing against machine and machine playing against machine. An important part is that the self-playing agent is trained by a Machine Learning algorithm and the entire learning process can be visualized.

3.2   Application Input and Output

The GUI expects from the user to select the game mode. If the user wants to play against the machine, he has the possibility to choose one of the trained agents as an opponent. If the user wishes to observe the training process of the agents, the application visualizes the entire learning process of the chosen agent.

3.3   Pong Game UML Class Diagram

Figure 2 shows the classes of the application and how they are related. The next subsections will show in detail contents and relations of the classes.

3.4   Main Window

The *MainWindow* class contains the main method created in this stage. The stage gets the respective scene by method call. All classes that create a scene use the method *insert()*. The *MainWindow* class also creates a file named *agents.agent* for the management of all agents.

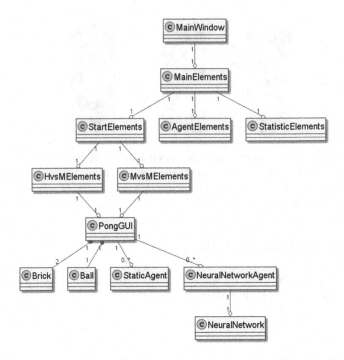

Fig. 2. Conceptual class diagram of the developed application.

3.5 Main Element

In this class the *MainElement* scene is created, which is passed to the class *MainWindow*. There are four buttons in this scene.

– *start:* Calls the *insert()* method of the class *StartElement*
– *agents:* Calls the *insert()* method of the class *AgentElement*.
– *statistics:* Calls the *insert()* method of the class *StatisticElement*.
– *exit:* terminate the program.

The Pong logo, as well as the background image, the size of the window and the size of the buttons, are passed to the class *MainElement* from the class *MainWindow*.

3.6 StartElement

In this class, the *StartElement* scene is created which is passed to the class *MainElement*. It appears by pressing the start button from the *MainElement* class. There are three buttons in this scene:

– *H vs. M* (Human vs. Machine): Calls the *insert()* method of the class *HvsMElement*.
– *M vs. M* (Machine vs. Machine): Calls the *insert()* method of the class *MvsMElement*.
– *back:* Calls the *insert()* method of the class *MainElement*.

3.7   AgentElements

In this class the *AgentElement* scene is created which is passed to the class *MainElement*. It appears by pressing the Agents button from the *MainElement* class. There are four buttons, one text field and two radio buttons put together in a button group in this scene:

- *new Agent:* By pressing on the new Agent button, the text field, the two radio buttons and the create button are added to the scene.
- *delete Agent:* Deletes the selected agent.
- *back:* Calls the *insert()* method of the class *MainElement*.
- *create:* Writes the name of the agent entered in the text field with the addition (slow) or (fast) in the file *agents.agent*. There are also two new files created. The files are named after the agent and the endings are *.diagram* and *.agent*.

3.8   StatisticElement

In this class, the *StatisticElement* scene is created which is passed to the class *MainElement*. It appears by pressing the *Statistic* button from the *MainElement* class. There are two buttons and a *ComboBox* in this scene:

- *comboBox:* to select the agent
- *show:* After pressing the show button, the file of the selected agent with the suffix *.diagram* is read. The scanned values are added to an *areaChart*, which is then added to the scene
- *back:* Calls the *insert()* method of the class *MainElement*.

3.9   HvsMElement and MvsMElement

In these classes the *HvsMElement* resp. the *MvsMElement* scene is created, which is called by the class *StartElement*. The scene appears by pressing the *H vs. M* resp. *M vs. M* button from the *StartElement* class. There are two buttons, and a *ComboBox* in this scene:

- *comboBox:* The *comboBox* is commited the *agentList* from the class *Main-Window*
- *start:* Creates a new object of type *PongGUI*, which is given the name of the selected agent
- *back:* Calls the *insert()* method of the class *StartElement*.

4   Neural Network

When creating the network, the edge weights are chosen randomly. Every time the *gameOver()* method is called, the position of the bricks and the position of the ball are checked. The distance between the ball and brick positions is the error passed to the *training()* method.

4.1 NeuralNetworkAgent

The *NeuralNetworkAgent* class is the link between the Pong game and the neural network. This class determines how many neurons and layers the neural network starts with. This network is created in the constructor of the class.

The method *learn()* takes the *Y* value of the ball and hands it over to the neural network, which outputs a value between 0 and 1. If the output is greater than 0.5 the brick will go up, if the output is lower than 0.5, the brick goes down. In this class, the *score* is also counted (how often the agent's brick hits the ball). The score is reset every time the agent's brick does not hit the ball. Each score is added to *scoreCounter* and saved at the end of the game in the file with the name of the agent and the proper suffix.

4.2 NeuralNetwork

The constructor of the neural network needs the following parameters:

- *int number_of_inputs:* specifies the number of input parameters the network expects.
- *int number_of_output:* specifies the number of outputs the network returns.
- *int number_of_layers:* specifies the number of layers the network has. There must be at least two layers (input-layer and output-layer). Each additional layer is a hidden layer.
- *int neurons_per_layer:* specifies the number of neurons each hidden layer should have.

The constructor creates the neural network by calling the method *network-Weights()* and thus sets the weights between the neurons. The weights are stored in a two-dimensional array *ArrayList <double[][]>* named *layer*.

networkWeights(): The method creates and returns a two-dimensional array of type *double[][]*. It expects two parameters (*int left_neurons, int right_neurons*). This array contains the weights between two layers. The value of the weights is chosen randomly at creation and can have a value between −1 and 1. The columns of the array represent which neuron is in the left layer and the rows represent which it is connected in the right one.

input(): This method performs calculations in the network. It returns the result of the output layer at the end. The *input()* method gradually gets the stored weight connections from the *ArrayList layer* and performs a matrix multiplication. The results of the hidden layer are stored in an *ArrayList <double[]>*. In the first run, the network uses the input values. For each additional multiplication it uses the results of the hidden layer.

matrixMultiplication(): This method performs a multiplication of two arrays. Before the result is returned, the array is again passed to the *sigmoidFunction()* method.

$$\begin{pmatrix} w_{11} & w_{21} \\ w_{12} & w_{22} \end{pmatrix} \cdot \begin{pmatrix} input1 \\ input2 \end{pmatrix} = \begin{pmatrix} w_{11} \cdot input1 + w_{21} \cdot input2 \\ w_{12} \cdot input1 + w_{22} \cdot input2 \end{pmatrix} \tag{2}$$

sigmoidFunction(): The method multiplies each value of the given array with the sigmoid function.

$$y = \frac{1}{1 + e^{-x}} \tag{3}$$

transpose(): The method takes a matrix as a two dimensional array and returns the matrix's transpose.

$$M = \begin{pmatrix} w_{11} & w_{21} \\ w_{12} & w_{22} \\ w_{13} & w_{23} \end{pmatrix} \quad M^T = \begin{pmatrix} w_{11} & w_{12} & w_{13} \\ w_{21} & w_{22} & w_{23} \end{pmatrix} \tag{4}$$

training(): This method updates the edge weights through an error backpropagation. The method expects a *double[]* array, which stores an error value for each output neuron. Each newly created matrix with updated edges is stored in an *ArrayList <double[][]>*. Since the hidden layer, which is still used as an input layer, becomes in the next run the output layer, the new error value for this layer must be calculated as follows:

$$error_{hidden} = \begin{pmatrix} \frac{w_{11}}{w_{11}+w_{21}} & \frac{w_{12}}{w_{12}+w_{22}} \\ \frac{w_{21}}{w_{11}+w_{21}} & \frac{w_{22}}{w_{12}+w_{22}} \end{pmatrix} \cdot \begin{pmatrix} e_1 \\ e_2 \end{pmatrix} \tag{5}$$

5 Experimental Results

5.1 Different Number of Genomes per Generation

Three Genomes per Generation: The default value of genomes created per generation was three. Different agents with three genomes were trained and the results were similar. Figure 3 shows a related learning behavior of the two agents.

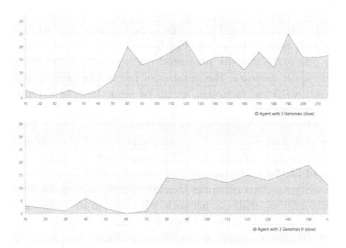

Fig. 3. Two agents with three genomes (*x*-axis: games played, *y*-axis: hits in one game).

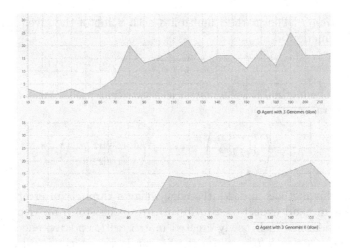

Fig. 4. Agent with five genomes compared to agent with three genomes (*x*-axis: games played, *y*-axis: hits in one game).

Five Genomes per Generation: It can be assumed that an agent with more genomes would automatically learn faster per generation than an agent with fewer genomes. After the agents were trained with three genomes, agents were then trained with five genomes per generation. Watching the new agent game, it is clear that the agent learned to follow the ball at an early stage, however, this behavior was not often repeated. This particular trend is barely evident in the diagram (Fig. 4).

5.2 Different Number of Hidden Layers and Neurons per Hidden Layer

In the following, different neural networks are compared with each other, all of them learning with backpropagation (Fig. 5).

The upper one is the already presented network with 1 hidden layer of two neurons. In the lower network, the number of layers has been increased. It has two hidden layers with two neurons. It can be seen that the extended network learns much earlier (Fig. 6).

Upper network with 2 hidden layers, each with two neurons. Lower network a hidden layer with four neurons. Both networks learn faster than the network with a hidden layer with two neurons. The upper network has a much steeper learning curve while the lower one starts learning earlier.

In the following, the two networks (backpropagation, evolutionary) are compared. Both algorithms utilize a network with two input neurons (*y*-value of the bricks, *y*-value of the ball), a hidden layer with two neurons and an output neuron. The evolutionary algorithm uses three genomes per generation.

It can be seen that the backpropagation algorithm learns slightly faster from the very beginning. From the 13th game a more clearly visible difference can be

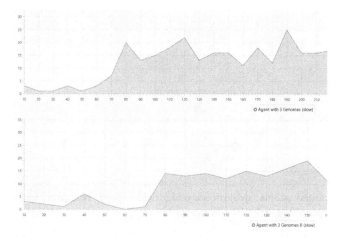

Fig. 5. Two agents. Both learn through backpropagation (*x*-axis: games played from 10 to 90, *y*-axis: hits in one game from 0 to 50).

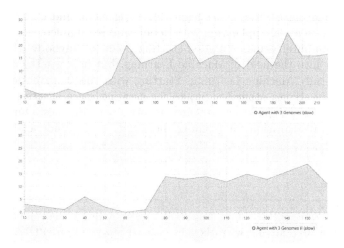

Fig. 6. Two agents. Both learn through backpropagation (top: x-axis: games played from 10 to 90, *y*-axis: hits in one game from 0 to 30, button: *x*-axis: games played from 10 to 80, *y*-axis: hits in one game from 0 to 30).

observed. It could be explained by the fact that the agent learns the backpropagation method with each call of the *gameOver()*-method, while the evolutionary algorithm has to work through all the genomes before creating a new generation exhibiting training progress.

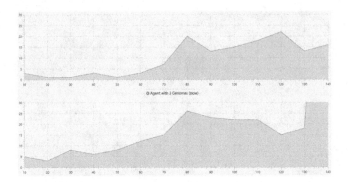

Fig. 7. Two trained agents: evolutionary algorithm (top), backpropagation (bottom) (x-axis: games played, y-axis: hits in one game).

6   Conclusion and Future Work

In this study, we have presented different Machine Learning approaches to the problem of creating a self-playing agent for Pong game and outlined how to implement it into a real application with details and real implementation. Pong, as a simple computer game, is a good starting point for anybody who has just taken interest into the field of Machine Learning, but it is not the end goal in the path to study Machine Learning. Starting with this contribution, we will continue to investigate other techniques for creating more effective agents for the Pong game and expand our research to other Atari games. Our goal is to devise a general algorithm for training the agent, that is independent of any input game (Fig. 7).

References

1. Bellis, M., Atari: History of the Entertaining Atari Video System and Game Computer, April 2017. https://www.thoughtco.com/history-of-atari-1991225. Accessed 11 June 2018
2. Brian, R.D.: Pattern Recognition and Neural Networks. Cambridge University Press, Cambridge (1996)
3. Draper, N.R., Smith, H.: Applied Regression Analysis, 3rd edn. Wiley, New York (1998)
4. Holland, J.H.: Adaptation in Natural and Artificial Systems. University of Michigan Press, Ann Arbor (1975)
5. Lunardi, A.C., et al.: Neural Network for Multitask Learning Applied in Electronics Games. In: European GAME-ON Conference (2013)
6. Mnih, V., et al.: Human-level control through deep reinforcement learning. Nature **518**(7540), 529 (2015)
7. Nork, B.: Pong Game with a Neural Network Agent. https://github.com/BeneNork/Pong-Game-with-a-Neural-Network-Agent. Accessed 11 June 2018

8. Parthasarathy, D.: Write an AI to Win at Pong from Scratch with Reinforcement Learning, September 2016. https://medium.com/@dhruvp/how-to-write-a-neural-network-to-play-pong-from-scratch-956b57d4f6e0. Accessed 11 June 2018
9. Rashid, T.: Make your own Neural Network (2016). ISBN 978-1530826605
10. Rinaldi, F.: Neural Network Plays Pong, September 2017. https://github.com/fabiorino/NeuralNetwork-plays-Pong. Accessed 11 June 2018
11. Roth, G.: Machine Learning for Java Developers: Set Up a Machine Learning Algorithm and Develop Your First Prediction Function in Java, September 2017. https://www.javaworld.com/article/3224505/application-development/machine-learning-forjava-developers.html. Accessed 11 June 2018

Managing Congestion in Vehicular Networks Using Tabu Search

Muhammad Ishaq[1], Mazhar Hussain Malik[2], and Mehmet Emin Aydin[3($\boxtimes$)]

[1] Institute of Southern Punjab, Multan, Pakistan
muhammadishaqkhoja@gmail.com
[2] Department of Computing, Global College of Engineering and Technology,
Muscat, Oman
mazhar@gcet.edu.om
[3] Department of Computer Science and Creative Technology,
University of West of England, Bristol, UK
mehmet.aydin@uwe.ac.uk

Abstract. In this era of communication, exponentially growing networks bring a lot of challenges to address for smoother network functionalities. Among them is efficiency in handling packet traffic to avoid and control congestion. A particular case is applicable to Vehicular Ad-hoc Networks, which are known with unbalanced resource utilisation, communication overheads, high transmission delay and least transmission capacity. This paper aims to minimise the delay and jitter for enhancing the Quality of Service (QoS) in Vehicular Adhoc Networks (VANET) using tabu search algorithm with multi-channel allocation capability. We proposed a scheme that prioritises each message considering the basis of message type or its substances, such as crisis, reference point, and administration oriented etc., and uses tabu search for scheduling the transmission of queued messages in order to enhance the efficiency, security, and durability of VANET. A comprehensive simulation is conducted to validate the proposed scheme and to evaluate the performances in comparison with other state-of-the-art approaches.

Keywords: Congestion control · Tabu search
Dynamic message scheduling · Priorities assignment messages
VANETs highways and urban environment

1 Introduction

Vehicular Adhoc Networks (VANETs) are particular implementation of Mobile Adhoc Networks (MANETs), which involves many network qualities and characteristics of MANETs alongside novel attributes including frequent changes in topologies. Secure and reliable communication with MANETs (and subsequently VANETs) remains as one of the main challenges being addressed through optimisation; by minimising delay and maximising the throughput of the network. In particular case of VANETs, communication can easily be

© Springer Nature Switzerland AG 2018
E. Pimenidis and C. Jayne (Eds.): EANN 2018, CCIS 893, pp. 118–129, 2018.
https://doi.org/10.1007/978-3-319-98204-5_10

blocked/disrupted due to not-well controlled message traffic and the channels can be saturated by collisions in the networks. Highly dynamically changing circumstances of VANETs such as fast mobility and frequent changes in network topology make congestion control problem in VANETs very attractive for investigation.

The clients of Vehicular Networks can be aware of unfavorable circumstances caused by vehicular correspondences and transmitting the data about encompassing situations [1,2,7]. Congestion happens in the channels when these channels face a dense vehicular saturation that is contending to access the channel. The proportional increase in the bandwidth usage due to growing the vehicle density causes congestion and may lead to crashes. The inevitable consequences of congestion include delay in transmission and losses in packet delivery, which affects the QoS and the efficiency of the Vehicular Networks. Therefore, an efficient QoS for VANETs is crucially required and can only be obtained by implementing a reliable and more secure networks [4,6].

Congestion control problem can be solved through studying control parameters in VANETs. Prospective approaches include tuning the transmission rate or applying some control mechanism on communication and deciding the size of conflict window. These would also encompass checking Arbitration Inter-Frame Spacing (AIFS) or studying prioritisation strategies on beacons. Inefficient congestion control strategies escalate to emerging severe issues including high transmission delay, unbalanced resource utilisation, less efficient bandwidth utilisation, correspondence overhead, and processing overheads are some example problems that are faced during congestion management strategies [6]. This study focuses on optimising the congestion control strategy to help minimise the packet loss and delay in transmission through optimally prioritising messages in the queues and subsequently reduce the channel loads.

In this paper, saturating traffic circumstances of VANETs are considered in managing message traffic and congestion control via optimising message routes using tabu search as one of prominent metaheuristic approaches. Tabu search has a success-proven record in problem solving and is adaptable to be used for instant decision making [5,9]. In the rest of the paper, VANET architecture and messaging infrastructure is introduced in Sect. 2, related works are discussed in Sect. 3, details of proposed approach in solving congestion control problem in Sect. 4 and experimental results are demonstrated and discussed in Sect. 5 following by conclusions in Sect. 6.

2   System Model

The system model of this problem appears to be VANET architecture composing the main characteristics of a typical VANET implementation. There are three incorporating sub-domains of VANETs; in-vehicle domain, ad-hoc domain, and infrastructure space. In-vehicle domain requires every vehicle to be furnished with On-Board Units (OBUs) to exploit them during short range remote correspondence for security and non-critical safety transmissions. The combination of OBUs and Road side Units (RSUs) compose Ad-hoc domain, which is to deal with inter-vehicle correspondences, where OBUs transmission can be in the form of single-hop or multi-hop transmission [19]. Infrastructure domain consist of RSUs and Hotspots (HS) while OBUs can utilize mobile systems architecture. These architecture associated with a number of different networks such as GPRS, GSM, UMTS, or any other like WiMAX as VANET architecture is displayed in Fig. 1.

The goal for In-vehicle domain is to facilitate inter-vehicle communication to share information such as position and speed to prevent accidents. Depending on the technology, vehicles may simply receive a risk warning of an accident or vehicle itself can take actions such as slow down or breaking. In the case of only one vehicle is in the communication range then uni-cast communication will be established otherwise it become a broadcast for that particular region. In the case of Adhoc domain where vehicles are communicating with RSU or vice-versa. Vehicle to RSU have one to one communication which include information such as speed, velocity and load on road and from RSU to Vehicles can have multi-cast or broadcast communication which includes information such as load on network and delay factors in that particular region. Infrastructure domain establishes communication between different RSU with server using Internet. Communication between the RSU to server can be uni-cast, multi-cast or broadcast to pass information depending upon situation.

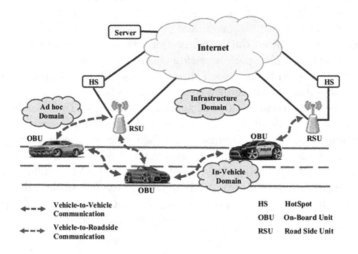

Fig. 1. VANET architecture

3   Related Works

Since a decade, a substantial amount of research has been done to handle congestion in VANETs through a variety of congestion control schemes, which appear to have significant limitations due to various shortcomings including frequent changes in topologies. In the following section, relevant recent research in this regard is overviewed.

Research reported in [10] addresses vehicular safety by introducing two congestion control approaches which work at MAC layer transmission queues including measurement and even driven detection. In case of event driven detection, congestion control is launched when safety message are generated. In order to detect congestion in measurement based mechanism, each device, after a specific time period, senses channel usage level to check over utilization. Once congestion is detected, MAC queue manipulation mechanism handle it. Carrier Sense Multiple Access with Collision Avoidance (CSMA/CA) is used in VANETs and it was defined by IEEE802.11p [8]. CSMA/CA uses default gateway strategy with exponential back-off mechanism to control congestion.

Rate adjustment scheme was proposed in [11] to ensure fair resource assignment in the networks. Mechanism measures traffic to compute channel utilization and help identify number of vehicles in the surrounding areas. However, this scheme leads to poor bandwidth management. Hannak demonstrates in [8] that the detected congestion control schemes categories as proactive and reactive. Reactive schemes obtain feedbacks from the networks to manage load, use channel states and react benchmarking with previously used pre-defined thresholds. On the other hand, proactive schemes are based on future prediction of traffic.

In [12,13], researchers investigate the impact of queue freezing schemes on VANETs performance, which are caused by stopping queue once event driven safety messages detected. VANETs interference was addressed by [14,15]. Authors of [18] introduce a mechanism to handle congestion in short range communication devices, while [16] reports window-based and rate-based congestion schemes. Window-based schemes use congestion windows at both sender and receiver sides, where the size of congestion windows is adjustable based on congestion state. On the other hand, rate-based schemes are established based on feedback algorithms and accept sending rate based on receiver side only [16].

In [17], Distributed Fair Power Adjustment scheme for VANETs (D-FPAV) was introduced, which proposes a dynamic control strategy for congestion control within transmission range. D-FPAV short beacon messages broad-casted periodically and provide vehicles position, direction and speed. In this scheme, receiving beacons from long range transmission are very low, while bandwidth is poorly managed due to greater beaconing and high traffic density.

An metaheuristic-based approach called Uni-Objective tabu search (UOTabu) is proposed in [7] to handle VANETs congestion, which offers performing calculations to determine channel usage level along with transmission range, which consequently leads to rate-tuning to minimise delay. This uni-objective scheme considers delay as an objective function. The study provides a comparison between CSMA/CA, D-FPAV and UOTabu schemes and suggests that UOTabu help decrease delay and increase throughput with the expense of higher computational complexities. The proposed scheme in this study does not consider the impact of emergency message broadcast on overall network performance and is verified with small size problems experimented only for time periods of 3 s.

Significance of studies can easily be judged based on above-mentioned literature, researchers implemented proactive scheduling based on future prediction of traffic as explained in [8], scheduling based on prediction is based on some probability of occurrence which effect QoS, while tabu search-based scheduling [7] implemented to address the delay, but the experimented time window is only 3 s, which does not allow the network to be congested with chosen problem sizes. Under such circumstances, tabu search would only incur more complexity overhead rather than help solve the congestion problem. On the other hand, the expected benefit of tabu search can only be realised in the cases of heavy-loaded networks running at least 1 h time.

4   Route Optimization Using Tabu Search

In this study, we decided to tackle the problem of heavily-loaded VANETS experimented for enduring much longer time periods using a bespoke tabu search approach. The implemented tabu search approach is to address congestion in VANETs considering both Urban and Highway environments. The main functionality expected of tabu search algorithm to help find optimal route, which resolves congestion problem of vehicular communication. This sort of dynamic routing/scheduling problems require rather instant and cannot afford long-elapsing decision processes. Hence, well-engineered decision processes such as local-search algorithms can be useful in searching for the best solution within such short time windows. Given these circumstances, existing state-of-art neighbourhood and local search approaches can be seen as candidates to solve these dynamic problems, however, there is a substantial risk of providing suboptimal solutions from the regions under consideration.

Tabu Search [3,9] is one of most mature metaheuristic approaches well-known with proven success [5], which offers facilitating local search rules in such a way that the best of neighbourhood is replaced with the current solution regardless of whether the solution is better or worse than the current one. As long as the best of the process is recorded, tabu search helps change the search regions systematically in this way. Moreover applied and issued moves within the search space are frozen for not to be re-considered for a particular while in order to prevent the search process from falling in vicious circle. The implemented algorithm is using memory structures to manage the visited solutions and if the potential solution has been previously visited within the short period of time or has violated a rule then it will be marked as tabu, then the algorithm will not consider this possibility repeatedly. The performance of proposed scheme is measured by calculating average delay, throughput, message generation rate and packet loss. Performance analysis is performed to compare it with other state of the art schemes.

This approach is implemented to comprise of two parts. The first one is the stage of priority allocation and the other is the stage of message scheduling. Priority allocation stage allocates priorities to the messages as in Fig. 2. As is shown, the complete mechanism works from application layer to physical layer.

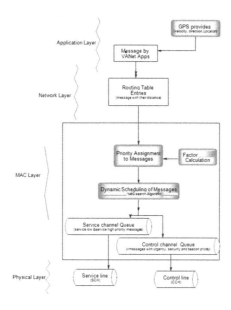

Fig. 2. Scheduling Procedure using Tabu Search with message priorities

In VANETs, Global Positing System (GPS) provides velocity, direction and location to the applications and then all the data moves to network layer. It is also part of the responsibilities of application layer to send broadcast messages to networks, which can include information, guidance and suggestions about routes such as information about upcoming congestion and suggestions according to that, average travel time to destination etc. Incoming messages can be of broadcast, uni-cast, or multi-cast. Routing table entries are created which shows message with their distances. In Medium Access Controller (MAC) layer, priorities are allocated to incoming messages based on calculated factors emerging last second. Tabu search algorithm transfer data to Service channel queue which include service low and high priority messages and to control channel queue which categories messages based on urgency, security and beacon priority. Finally messages coming from Service channel queue move to service line (SCH) and message from control channel queues move to Control line (CCH) at physical layer.

Tabu search algorithm implemented in this study starts with initialising a number of parameters; *message type, factors <delays, congestion, intersection>, the maximum number of iteration Tabu List* and *all counts* including *intensification.* The algorithm is set to run until the pre-set stopping criterion is met.

A typical problem state is a set of messages in the queue, denoted as $\mathcal{M}$, ordered based on associated priorities, $\mathcal{P}$. Hence, a set of messages to be transmitted can be denoted as $X = \{x_i | i = 0, .., |\mathcal{M}|\}$, where x_i denotes a typical message representing a pair of $< m_i, p_i >$ and m_i and p_i denote i_{th} message and its priority, respectively. A randomly generated solution will have an order based on the priorities. A neighbourhood function $f_N(X_i)$ takes an existing solution and randomly swaps any two messages, say x_j with x_l, if their priorities are equal or very close to each other, **iff** $p_j = p_l$ **OR** $|p_j - p_l| \leq \Theta$, where Θ is a threshold setup for differentiating priorities. A set of neighbours, $N = \{n_k | k = 0, .., |N|\}$, is generated using $f_N(.)$ to the pre-set neighbourhood size. A candidate set, $C = \{c_j | j = 0, .., |C|\}$, is filtered out of N with checking if any neighbour, n_k, is resulted of any tabu *move*. In each iteration, the best performing c_j is replaces the current solution, $X_i \rightarrow c_j$.

Once a neighbouring solution, n_k, is generated, the action taken, which is called *move* and denoted with $mv < x_i, x_j >$, is recorded in the short memory. Once X_i is replaced with c_j, then $mv < X_i, c_j >$ will be added to the tabu list, $T = \{t_m < X_m, c_m > | m = 0, .., |T|\}$, which is aged iteration-by-iteration, and released once iterated $|t|$ times. Tabu list is considered as the long-term memory in this optimisation process. The quality of each solution is measured through network simulations, which are selected to be average packet delay and average packet loss. This calculation is conducted once a particular state of the problem (solution) is passed into the network simulation in an opted data structure.

Following initialisation step, a typical tabu search iteration consists of neighbourhood and candidate list generations, selecting the best candidate as the new solution and admitting the move in tabu list, and finally updating the tabu status of the whole tabu list. There might be some conflicting cases, where none of the solutions in the neighbourhood is eligible to be candidate since all are included in tabu list. Such cases would trigger re-generating the neighbourhood. If successively generated neighbourhoods are not producing eligible solutions within a pre-set intensification period, which is deviced to remedy this sort of cases.

5   Performance Evaluation

In order to evaluate performance of proposed approach in comparison with existing commonly used approaches, network and mobility simulators, SUMO and NS-3, are used. Connection between SUMO and NS-3 is defined using MOVE, which is a mode used for generating mobility in VANETs. As an urban scenario, a six-lane road within urban environment is simulated. Table 1 presents the parametric details considered in the simulation, while 802.11p is taken into account as an interconnection standard. CSMA/CA approach is adopted as transferring approach in MAC layer, Nakagami and 2-dimensional grounds (models to show highway conditions) are preferred to design propagation into urban and highway conditions. The data traffic is assumed to conform the Poisson distribution.

5.1   Simulation Parameters and Environments

Performance of suggested approaches are examined with the state-of-the-art approaches known as Enhanced Distributed Channel Access (EDCA), First In First Out (FIFO) and dynamic scheduling. Across these performance evaluations, the following key performance indicators (KPI) are among those to be considered for examination purposes:-

– **Message lost count** during simulation run time, the quantity of messages lost is checked to find the ratio of loss.
– **Average delay in transmission** from dispatcher to receiver, average time needed for transferring messages.
– **Ratio of message loss** as the quantity of messages lost in comparison with no of transferred messages.
– **Average throughput** within unit of time, which is the ratio of efficiently received messages on interconnection channels.
– **Waiting time in queue**, before de-queuing and transmitting towards channels, average waiting time for messages in SCH and CCH queues.

As may be seen, two main performance areas are identified; *delay* and *packet loss*, which have been chosen as the two main KPIs in this study. The following evaluations and discussions have been put together on the basis of these two in averaged form. All the calculations are conducted to determine these two KPI values through specifically configured network simulation with the parametric details provided in Table 1.

Table 1. Parameters to configure simulation for highway* and urban** cases

Parameters	Value	Parameters	Value
MAC type	802.11p	Bandwidth	75 MHz
Frequency	5.850 GHz to 5.925	Simulation time	200 s to 1 h
Message size	Emergency (578 bytes) beacon (500 bytes)	Speed of vehicles	80–120 km/h* 0–50 km/h**
Number of lanes	4 (2 in every position)** 6 (3)*	Total road length	5400 m* 1600 m × 1700 m**
Number of vehicles	400, 300, 200, 150, 100, 50	Transmission rate	10 Mbps

5.2 Comparative Analysis of Proposed Algorithms with Others

As per comparative analysis, tabu search-based scheduling (St-Sch) approach, which will be named tabu Scheduling here-forth, is observed in general that it helps reduce the delay of transmission while FIFO does not work effectively in high quality transport environments since it de-queues the packets with none prioritised ones. EDCA prioritisation mechanism fails to manage congestion when there are large numbers of priority messages that escalates to high delay for low priority data.

As indicated in Fig. 3a the average delay in cases of highways by the proposed scheme is 4 ms when there are 200 vehicles, while the best of the state of the art techniques considered, so-called dynamic scheduling, produces much higher delay starting form 5.5 ms. On the other hand, FIFO and EDCA mechanisms are much less efficient as delay increases exponentially with growing number of vehicles. Figure 3b shows the packet loss ratio on different load of vehicles.

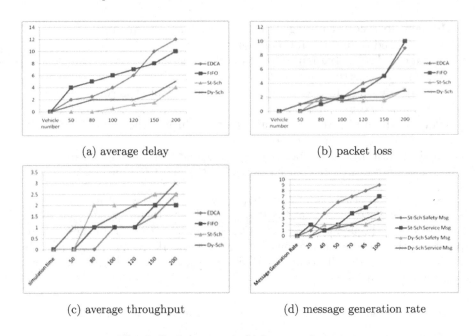

(a) average delay

(b) packet loss

(c) average throughput

(d) message generation rate

Fig. 3. Performances in highway environments

Studies suggest that dynamic scheduling (Dy-Sch) and tabu scheduling (St-Sch) methods perform better than other rival methods.

Performance analysis is extended with cases in urban environment alongside the highway cases. The results are plotted as in Fig. 4a and b, where it is clearly observed that EDCA and FIFO provide the worst performance with growing number of vehicles, especially beyond 120, while both dynamic scheduling and the proposed approach, tabu scheduling, perform very similarly noting that the measure of average packet loss by tabu and dynamic scheduling approaches are much less, in some cases the performance of dynamic scheduling is little bit higher.

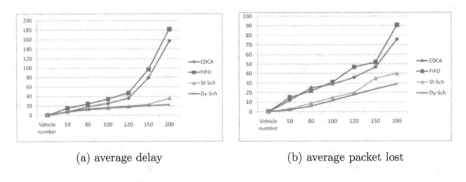

(a) average delay

(b) average packet lost

Fig. 4. Performances in urban environment in vehicle count

Figure 3c suggests that the average throughput by tabu scheduling is higher and offer high stability in comparison to rival approaches, similarly while comparing safety and services message with tabu and dynamic search, it is clearly shown in Fig. 3d that message generation rate of tabu scheduling is much higher assuring a higher level of security and efficient processing of service messages.

Table 2. Average delay in urban and highway environments with growing number of vehicles

Number of vehicles	Urban				Highway			
	EDCA	FIFO	Dy-Sch	St-Sch	EDCA	FIFO	Dy-Sch	St-Sch
100	25	34	15	14	4	6	2	0.5
200	157	182	25	22	23	10	5	4
300	300	510	48	31	30	25	10	7
400	501	780	80	39	50	40	19	10

Table 2 presents the average delay scores in transmission by each scheme, comparatively, in high load circumstances, the performance of the schemes are measured with different loads from 100 to 400 vehicles in cases of both urban and highway infrastructures simulated for 1 h. It can be observed that the average delay in urban cases are much higher than highways, as expected. It is also clear that both dynamic scheduling and tabu scheduling schemes significantly help reduce the average delay beyond 300 vehicles, in both environmental circumstances, while tabu scheduling clearly achieves better than dynamic scheduling. Average delay caused by tabu scheduling remains as 31 in urban and 7 in highway environments when the network size is 300 vehicles, it become 39 and 10 when the network is of 400 vehicles.

Table 3 shows average packet loss score by each scheme comparatively, where packet loss measure is much lower in highway circumstances as expected. The network's workload is set to high level, as is in delay cases, spanning from 100

Table 3. Average packet-loss in urban and highway environments with growing number of vehicles

Number of vehicles	Urban				Highway			
	EDCA	FIFO	Dy-Sch	St-Sch	EDCA	FIFO	Dy-Sch	St-Sch
100	29	31	12	9	2	2	1.5	1.5
200	76	91	39	18	9	10	3	3
300	136	216	56	25	17	23	10	5
400	229	481	86	31	39	49	19	8.5

to 400 vehicles. It is observed that both dynamic and tabu scheduling schemes remain significantly lower than the other two classical schemes, especially beyond 300 vehicles. Here, the score of tabu is 25 losses in urban and 5 in highway environments under the circumstances of the network of 300 vehicles, while 31 and 8.5 achieved for a network of 400 vehicles. This clarifies and confirms the usefulness of tabu search in solving these problems.

6   Conclusion

In this paper, a novel routing/scheduling approach called tabu scheduling is studied for handling congestion control and message traffic in highly dynamic networks such as Vehicle Ad-hoc Networks (VANET). The proposed technique is experimented and evaluated in comparison with two classical (FIFO, EDCA) and one most recently introduced state-of the-art approaches. It is observed that the proposed approach helps handle congestion much more efficiently for both urban and highway environments. Tabu scheduling results are highly admirable as it focuses on transmission delay and message superiorities, on the other hand, dynamic approach is only focusing on superiorities messages. Results show that tabu Scheduling rank first than dynamic scheduling while EDCA and FIFO are least efficient. Thus, within VANETs, reliable and secure atmosphere provided through tabu scheduling.

References

1. Golestan, K., Jundi, A., Nassar, L., Sattar, F., Karray, F., Kamel, M., Boumaiza, S.: Vehicular Ad-hoc Networks (VANETs): capabilities, challenges in information gathering and data fusion. In: Kamel, M., Karray, F., Hagras, H. (eds.) AIS 2012. LNCS, pp. 34–41. Springer, Heidelberg (2012). https://doi.org/10.1007/978-3-642-31368-4_5
2. Golestan, K., Sattar, F., Karray, F., Kamel, M., Seifzadeh, S.: Localization in vehicular ad hoc networks using data fusion and V2V communication. Comput. Commun. **71**, 61–72 (2015)
3. Barbarosoglu, G., Ozgur, D.: A tabu search algorithm for the vehicle routing problem. Comput. Oper. Res. **26**(3), 255–270 (1999)
4. Ghosh, T., Mitra, S.: Congestion control by dynamic sharing of bandwidth among vehicles in VANET. In: 2012 12th International Conference on Intelligent Systems Design and Applications (ISDA), pp. 291–296. IEEE, November 2012
5. Guan, C.H., Cao, Y., Shi, J.: Tabu search algorithm for solving the vehicle routing problem. In: Third International Symposium on Information Processing, Qingdao, pp. 74–77 (2010)
6. Sepulcre, M., Gozalvez, J., Altintas, O., Kremo, H.: Integration of congestion and awareness control in vehicular networks. Ad Hoc Netw. **37**, 29–43 (2016)
7. Taherkhani, N., Pierre, S.: Congestion control in vehicular ad hoc networks using meta-heuristic techniques. In: Proceedings of the 2nd ACM International Symposium on Design and Analysis of Intelligent Vehicular Networks and Applications, pp. 47–54. ACM, October 2012

8. Hannk, G.: Congestion and awareness control in cooperative vehicular networks. In: Mobile Communications Seminar SS 2013 (2013)
9. Ho, S.C., Haugland, D.: A tabu search heuristic for the vehicle routing problem with time windows and split deliveries. Comput. Oper. Res. **31**(12), 1947–1964 (2004)
10. Zang, Y., Stibor, L., Cheng, X., Reumerman, H.-J., Paruzel, A., Barroso, A.: Congestion control in wireless networks for vehicular safety applications. In: Proceedings of the 8th European Wireless Conference (2007)
11. Baldessari, R., Scanferla, D., Le, L., Zhang, W., Festag, A.: Joining forces for VANETS: a combined transmit power and rate control algorithm. In: 7th International Workshop on Intelligent Transportation (WIT), Hamberg, Germany, p. 15 (2010)
12. Mughal, B.M., Wagan, A.A., Hasbullah, H.: Efficient congestion control in VANET for safety messaging. In: 2010 International Symposium in Information Technology (ITSim), Kuala Lumpur, pp. 654–659 (2010)
13. Bouassida, M.S., Shawky, M.: A cooperative congestion control approach within VANETs: formal verification and performance evaluation. EURASIP J. Wirel. Commun. **2010**, 11 (2010)
14. Le, L., Baldessari, R., Salvador, P., Festag, A., Zhang, W.: Performance evaluation of beacon congestion control algorithms for VANETs. In: 2011 IEEE Global Telecommunications Conference (GLOBECOM 2011), Houston, TX, USA, p. 16 (2011)
15. Bratko, I., Michalski, R.S., Kubat, M.: Machine Learning and Data Mining: Methods and Applications. Wiley, London (1999)
16. Subramani, B., Chandra, E.: A survey on congestion control. Global J. Comput. Sci. Technol. **9**, 82–86 (2010)
17. Torrent-Moreno, M., Santi, P., Hartenstein, H.: Distributed fair transmit power adjustment for vehicular ad hoc networks, In: 2006 3rd Annual IEEE Communications Society on Sensor and Ad hoc Communications and Networks, SECON 2006, Reston, VA, pp. 479–488 (2006)
18. He, J., Chen, H.-H., Chen, T.M., Cheng, W.: Adaptive congestion control for DSRC vehicle networks. IEEE Commun. Lett. **14**, 127–129 (2010)
19. Lu, N., Zhang, N., Cheng, N., Shen, X., Mark, J.W., Bai, F.: Vehicles meet infrastructure: toward capacitycost tradeoffs for vehicular access networks. IEEE Trans. Intell. Transp. Syst. **14**(3), 1266–1277 (2013)

Network Intrusion Detection on Apache Spark with Machine Learning Algorithms

Elif Merve Kurt[1,2(✉)] and Yaşar Becerikli[1,3(✉)]

[1] Computer Engineering Department, Kocaeli University, Umuttepe Campus,
41380 Kocaeli, Turkey
ybecerikli@kocaeli.edu.tr
[2] Ctech Information Technology Incorporated Company, Teknopark Istanbul,
Istanbul, Turkey
elif.merve.kurt@ctech.com.tr
[3] Forensic Computing Department, Ankara Group Presidency,
Council of Forensic Medicine, Ankara, Turkey

Abstract. The continuous increase in internet-based services makes network traffic data larger and more complex day by day. This makes it increasingly difficult to detect network attacks, and therefore requires more efficient and faster data processing methods to ensure network security. For this purpose, many intrusion detection systems have been developed and development works are continuing.

This study; by comparing the performance of machine learning algorithms on the same network data, aims to establish a reference source for the developed intrusion detection systems. In this study; all data of KDD Cup'99 were run on Logistic Regression, Support Vector Machine, Naive Bayes and Random Forest from machine learning algorithms using Apache Spark a big data technology; and the results were analyzed comparatively.

Keywords: Network attack · Intrusion detection system · KDD Cup'99
Big data · Apache spark · Machine learning

1 Introduction

The advancement in computer technologies and the acceleration of computer networks make the development of internet based services compulsory. With the development of internet-based services, network traffic data became increasingly complex in terms of management. This makes network attacks difficult to detect and concurrently threatens network security. Although more effective methods of data processing are developed to protect against threats, attackers are constantly creating new attacks. This cyclical situation makes network attacks the focus of researchers.

Intrusion detection systems, which have been in continuous development since the 1980 are used in combination with various learning algorithms; they can be examined in two main sections: Information Based and Behavioral Based [1]. The use of pattern classification in detecting attacks is applied in various fields. Attacks can be better captured by using pattern classification, information and behavior based system logic. There are two main methods [2]: Supervised Learning and Unsupervised Learning.

E. Pimenidis and C. Jayne (Eds.): EANN 2018, CCIS 893, pp. 130–141, 2018.
https://doi.org/10.1007/978-3-319-98204-5_11

In this study; KDD Cup'99 was selected as the experimental data. For faster data processing, big data technology Apache Spark was used. Information based intrusion detection system has been established using Logistic Regression, Support Vector Machine, Naive Bayes and Random Forest machine learning algorithms. The output of this study is performance analysis of intrusion detection systems developed with different machine learning algorithms. The performed comparative analysis will be a reference source for researchers. In the first section of the study, the purpose and the subject are explained; in the second part, related works are mentioned. In the following sections; the system architecture, technologies dataset and algorithms used in system, discussion, conclusion and references are included.

2 Related Works

With the increase in network traffic data, it has become increasingly difficult to detect network attacks. Resolving this problem with faster and more effective methods has become the main focus of researchers [3–7].

In literature studies vastly use Apache Spark as a big data processing tool because of its ability to rapidly analyze network traffic data and capture attack traffic [3, 5, 7]. It is emphasized by researchers that the Apache Spark is a fairly suitable tool for machine learning algorithms that requires repetition [5].

In literature, feature extraction algorithms well known for extracting qualified features have been utilized. These qualified features have been given as the dataset to the network intrusion detection system in which classification based methods are used [3, 4]. The mechanism of the system is as follows; after the information of the available network data is learned by the system, the learned information of the new incoming network data is classified as either normal or attack [6].

Overall, well known machine learning algorithms Logistic Regression [3, 4, 6], Support Vector Machine [3, 6], Random Forest [3, 6], Gradient Boosted Decision Tree Algorithms [3, 4], Naive Bayes [3] and Decision Tree Algorithms [6] are frequently used as a classification method in network intrusion detection system studies.

The KDD Cup'99 dataset which consists of data from DARPA (which is a real time dataset) was used as an experimental dataset for performance comparison verification for classification based methods [3, 6, 7]. This experimental dataset has become a standard in literature. Forest Cover Type and Internet advertising data are also among the experimental datasets used [5].

In this study, the Apache Spark Machine Learning Library was used for attack detection from big network traffic. Unlike similar studies, the algorithm extensions and data structures that came with the latest release of Apache Spark are used. Machine learning algorithms Logistic Regression, Support Vector Machine, Naive Bayes and Random Forest which are proved to be suitable for network data structure have been used in the system to enable comparability with previous studies. For the same purpose, speed and accuracy tests were performed by changing the parameters on the machine learning algorithms using the all KDD Cup'99 dataset. The algorithms are reported with parameters, giving the most optimal results in terms of time and accuracy.

In other studies, more efficient intrusion detection systems were tried to be established in terms of speed and accuracy by operations such as feature selection performed by various algorithms used on experimental dataset. The purpose of this study is not to create a more efficient and faster system, but rather to reveal the development that has occurred in the existing systems. The results clearly show how effective the development of Apache Spark tool is in detecting attacks on big network data [3].

3   System Architecture

The architecture of the developed system consists of training and prediction stages in general (see Fig. 1). The system is expected to detect the learned attacks.

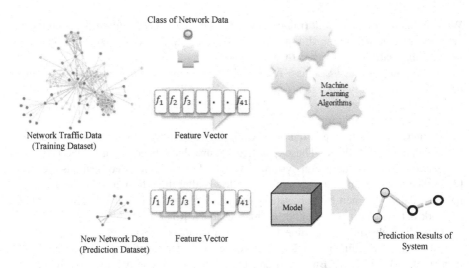

Fig. 1. Network intrusion detection system architecture

The training dataset is applied to Logistic Regression, Support Vector Machine, Naive Bayes, Random Forest machine learning algorithms. A model for each algorithm is created. Prediction dataset without the class label is given to the generated models and the predicted class outputs of each algorithm are obtained. The accuracy of the algorithm is calculated by the prediction outputs. The processes performed while the system is being developed are: data preprocessing, training and prediction.

Data Preprocessing: The dataset is separated as the feature vector generated by extracting the features that will be effective in identifying the problem and the class label (There is no class label in the prediction dataset). Categorical features are converted to numerical features. Available dataset are separated in a ratio of 60% training dataset and 40% prediction dataset. The training data consists class label; however the prediction dataset does not.

Training Phase: is the step where the machine is being trained. At this step; it is extremely important to use enough, quality data and to select the algorithm that is suitable for the problem. The machine builds a model of itself with the knowledge obtained from the presented data and the machine learning algorithm rules determined for its use. In other words, all the knowledge that the machine knows about the type of data that will help to predict the new incoming data class is called the model. The mentioned model structure is the output of the training phase.

Prediction Phase: When the machine is compared with an unlabeled data, it begins to work on the model it creates using the algorithm rules determined for that data type and tries to predict the class of the new data with the knowledge previously learned. The output of the prediction phase is the class label for which the machine has predicted from the new data.

4 Technologies, Dataset and Algorithms Used in System

4.1 Big Data Technology: Apache Spark

Apache Spark; is an open source code platform written in Scala language, designed for big data processing, allowing parallel processing on datasets. Compared with another big data processing technology, Hadoop consists of two components MapReduce and HDFS (Hadoop Distributed File System). It is a MapReduce alternative that provides big data parallel processing. Compared to MapReduce it performs memory operations 100 times faster [8]. Apache Spark supports many programming languages. It provides an easy to use API for Scala, Python, Java, and SQL. Also, with the included machine learning library MLlib, machine learning algorithms can run on more than one machine, so that faster analysis can be performed on big data.

4.2 Dataset: KDD Cup'99

KDD Cup'99 is a dataset used in the "Third International Knowledge Discovery and Data Mining Tools" organization to create a network intrusion detector. It includes a standard set of network data that encompasses a wide range of fraudulent attacks on military network environments. The dataset was prepared by Stolfo et al. in 1999 and has become the most widely used dataset for the evaluation of anomaly detection. The KDD Cup'99 training dataset contains approximately 4.900.000 single link vectors. A link vector consist of 41 features and is labeled as normal or special attack type (the dataset contains 23 types of link types in total). It is believed that newly developed attacks can be detected with the knowledge learned from the registered attacks. According to researchers, established information is enough for the detection of unknowns.

Attacks Types Contained Within: Attack types in the KDD Cup'99 dataset can be basically examined in four main sections [9]:

- **DoS-Denial of Service Attack:** In this type of attack; the attacker creates some very busy or very busy memory resources by doing some calculations to handle logical requests; denies the users access to the machine.
- **U2R-User to Root Attack:** These attacks are a type of exploitation made by deciphering passwords or social engineering. An attacker could exploit some vulnerabilities to gain root access to the system; starts attacking by accessing a normal user account on the system.
- **R2L-Remote to Local Attack:** An attacker who exploits some security vulnerabilities to provide local access, attack to target machine by send a packet over the network to the machine like a normal user.
- **Probing Attack:** It is an attempt to gather information about the computer network for a specific purpose by passing security measures (Table 1).

Table 1. KDD Cup'99 attack types

Dos	Probe	R2L	U2R
smurf	portsweep	ftp_write	buffer_overflow
teardrop	ipsweep	guess_passwd	perl
neptune	satan	imap	loadmodule
back	nmap	multihop	rootkit
pod		phf	
land		spy	
		warezclient	
		warezmaster	

Features Contained Within: Features in the KDD Cup'99 dataset can be basically examined in three main sections [9]:

- **Basic features:** This category covers all the features available from the TCP/IP connection. Many of these features are significant delay reason in attack detection.
- **Traffic features:** These features are calculated according to a window time interval and are examined in two groups:
  - **Same host:** It only examines the connections within the last 2 s that have the same destination host with the instant connection and it computes statistics about protocol behaviors such as service.
  - **Same service:** It only examines the connections within the last 2 s that have the same service with the instant connection.
- **Content features:** Unlike most DoS and Probing attacks, R2L and U2R attacks do not have frequently repetitive pattern sequences. While DoS and Probing attacks contain many connections in very short time periods, R2L and U2R attacks are embedded in the data part of the package and normally only contain a single connection. When detecting such attack types, some features are needed to look at suspicious behavior in data fragments (such as failed logins). These are called content features (Table 2).

Table 2. KDD Cup'99 attributes

No	Attribute	No	Attribute	No	Attribute
1	duration lenght	15	lsu_attempted	29	same_srv_rate
2	protocol_type	16	lnum_root	30	diff_srv_rate
3	service	17	lnum_file_creations	31	srv_diff_host_rate
4	flag	18	lnum_shells	32	dst_host_count
5	src_bytes	19	lnum_access_files	33	dst_host_srv_count
6	dst_bytes	20	lnum_outbound_cmds	34	dst_host_same_srv_rate
7	land	21	is_host_login	35	dst_host_diff_srv_rate
8	wrong_fragment	22	is_guest_login	36	dst_host_same_src_port_rate
9	urgent	23	count	37	dst_host_srv_diff_host_rate
10	hot	24	srv_count	38	dst_host_serror_rate
11	num_failed_logins	25	serror_rate	39	dst_host_srv_serror_rate
12	logged_in	26	srv_serror_rate	40	dst_host_rerror_rate
13	lnum_compromised	27	rerror_rate	41	dst_host_srv_rerror_rate
14	lroot_shell	28	srv_rerror_rate		

4.3 Machine Learning Algorithms

Machine learning algorithms using specific parameters determine the decision rules which carry the correct decision making feature on the new data. They object to find the most suitable model that characterizes the data best, because the better the model characterizes the data, the better decision-making mechanism works.

4.3.1 Logistic Regression

Logistic regression is used for problems whose dependent variable is categorical. The purpose is to create a model of the relation between dependent and independent variables [10]. The ease of mathematical interpretation of the model makes this method eligible. The independent variable is "x", the dependent variable is "Y" and the conditional average of the dependent variable is "$\pi(x) = E(Y|x)$". Logistic regression model function is shown in Eq. (1). If the category number of Y is 2, the conditional average value range will be "$0 \leq E(Y|x) \leq 1$".

$$\pi(x) = E(Y|x) = \frac{e^{\beta_0 + \beta_1 x}}{1 + e^{\beta_0 + \beta_1 x}} \tag{1}$$

The likelihood of encountering the dependent variable is "$P(Y = 1|x) = \pi(x)$" and not encountering the dependent variable is "$P(Y = 0|x) = 1 - \pi(x)$". The conditional average must be transformed to logit to take the form of linear expression. This transformation is displayed at Eq. (2).

$$g(x) = ln\left[\frac{\pi(x)}{1 - \pi(x)}\right] = ln\left[\frac{P(Y = 1|x)}{P(Y = 0|x)}\right] = \beta_0 + \beta_1 x \tag{2}$$

When the category number of dependent variable Y is larger than 2, multinomial logistic regression will step in. Multiclass classification will be carried out by combining binary classifier. For instance if Y has {0, 1, 2} categories and Y = 0 is the reference category, two logistic model will be such that "Y = 0 against Y = 1" and, "Y = 0 against Y = 2". This method is known as one-vs-all.

4.3.2   Support Vector Machine

Support Vector Machine (SVM) is one of the most effective machine learning algorithms. It is frequently used for solving complex classification problems. Although it was initially designed to classify two classes of linearly separable data, today it is used to classify data that are composed of more than two classes which cannot be separated linearly. SVM is based on the prediction of the hyper plane, which is the decision function that will determine the data classes [11].

Assume the dataset that can be linearly separated is expressed as $(x_1, y_1), (x_2, y_2)$, ..., (x_n, y_n). n is the total number of classes in the dataset. As $y_i \in \{+1, -1\}$, y_i values hold the class label of x_i values (i = 1, ..., n) (see Fig. 2) [12]. There are many hyper planes that can separate the dataset. The aim here is to find a hyper plane that maximizes the distance between the closest points separating the two classes. H_0 is the optimal hyper plane. H_1 and H_2, known as support vectors, are the vectors that determine the decision boundary width. H_1 and H_2, are obtained by Eq. (3) and the optimal hyper plane H_0 is obtained by Eqs. (4) and (5) [13]. In these equations, "w" is weight vector and "b" is bias.

$$|w.x_i + b| = 1 \tag{3}$$

$$w.x_i + b \geq 1, \quad \forall x_i \in (y = +1) \tag{4}$$

$$w.x_i + b \leq 1, \quad \forall x_i \in (y = -1) \tag{5}$$

For binary classification problems using the dataset that can be linearly separated, the SVM decision function obtained as a result of a number of optimization operations is shown in Eq. (6) [13]. "λ" is the Lagrange multiplier and (*) is the expression of the optimum values.

$$f(x) = sign\left(\sum_{i=1}^{n} y_i \lambda_i^* (x.x_i) + b^*\right) \tag{6}$$

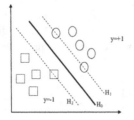

Fig. 2. Linearly Separable Dataset

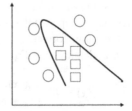

Fig. 3. Not Linearly Separable Dataset

The current dataset may not be linearly separated as in Fig. 3. The solution in such classification problems is to move the data to a larger dimensional space called the feature space. In the new space, each x variable is "$\phi(x)$" feature vector that is expressed as "$\phi(x) = \alpha_1\phi_1(x), \alpha_2\phi_2(x), \ldots, \alpha_n\phi_n(x)$". After changing the data space, the decision function is obtained by using kernel functions expressed as "$K(x, y) = \phi(x).\phi(y)$". The obtained decision function is shown in Eq. (7). Kernel functions can be linear, polynomial or radial basis [13].

$$f(x) = sign\left(\sum_{i=1}^{n} y_i\lambda_i^* K(x, x_i) + b^*\right) \tag{7}$$

In multi-class SVM problems, datasets are classified using more than one binary SVM. There are many methods recommended for multi-class SVM problems, the most important ones being one-vs-one and one-vs-all approaches [14].

4.3.3 Naive Bayes

Naive Bayes is a probability-based, widely used method. It assumes that each feature is independent. Modeling is easy and works better with big datasets. The learning model is created based on the probabilities of the dataset belonging to classes [15]. Naive Bayes produces the posterior probability $P(y|x)$, by using $P(x|y)$ and $P(y)$ priori probabilities shown in Eq. (8) which is called the Bayesian rule.

$$P(y|x) = \frac{P(x|y).P(y)}{P(x)} \tag{8}$$

$$P(y|x) = P(x_1|y).P(x_2|y).P(x_3|y)\ldots P(x_n|y).P(y) \tag{9}$$

Events that are independent of the probability of occurrence with Bayes are examined. "$x_1, x_2, \ldots, x_n$" in Eq. (9) show independent events. The posterior probability is computed for each class y. The class that produces the max value is determined as the class label of data.

4.3.4 Random Forest

It is one of the widely used methods of collective classification. In this method, multiple classifiers are created instead of a single classifier. The mentioned classifiers are tree type classifiers. Instead of producing a single decision tree, the decisions of a multitude of multi variable decision trees trained with different training dataset are all combined. Therefore, it is referred to by researchers as the "collection of tree-type classifiers" [16, 17]. It surpasses other methods of collective classification in terms of both speed and accuracy. Its competence and correctness makes it a practical classifier [18]. Steps to create a Random Forest model are:

(1) With the selections made through the training dataset, a new training data is generated. T is the training dataset, T_k is the new training data generated for the tree to be created. Two-thirds of T_k will be bootstrapping samples for the decision tree. The remaining 1/3 will be used as OOB (Out of bag) data. When the tree is created, the classifier will be tested using OOB data and errors will be computed.

(2) Two parameters must be defined by the user before the algorithm starts to operate [19]: "m", is the number of variables used in each node to perform best partitioning in the tree structure. When the value of m is chosen as the square root of the total number of variables, the best results are achieved [20]. "N" is the number of tree generations that is required. In researches, in most cases 500 trees are sufficient [23].

(3) From the new training data, a decision tree is created by random property selection method. Pruning is not carried out in trees [21, 22]. The algorithm becomes more effective classifier when not pruned compared to other tree type classifiers. The Random Forest algorithm uses the Classification and Regression (CART) algorithm to generate a decision tree [22]. The CART algorithm uses particular partitioning algorithms when deciding tree nodes. The Random Forest uses "gini" from these partitioning algorithms [17]. The Random Forest algorithm process steps are repeated until N value which is defined by the user.

At the end of each process cycle, the generated classifier is tested with the initially allocated OOB data and the OOB error is obtained. This error value shows the effect of the variables used in the classifier. While predicting a class belonging to a new incoming data in the random forest algorithm, first the new data is assigned to each node in the forest generated during the modeling phase. Then the class outputs produced by each tree are recorded. The class that receives the most votes is determined as the class of the new incoming data.

5   Discussion

The developed software was run on an Ubuntu operating system. The developments were made using the DataFrame based API of Apache Spark Machine Learning Library (Mllib).

The properties of Apache Spark which is configured with 4 GB executer and driver memory is as follows:

- Spark version: 2.2.1 (Dec 01 2017)
- Package type: Pre-built for Apache Hadoop 2.7 and later

The dataset used is the KDD Cup'99 dataset that contains in total "4898431" network link data. The data were used as training and prediction dataset by a separation ratio of 60%–40%.

Logistic Regression was applied using the one-vs-all method with default over fitting parameters. *Support Vector Machine* was implemented using default over fitting parameters and one-vs-all method with linear kernel function. *Naive Bayes* is implemented using default parameters. *Random Forest* parameters used are as folows; the number of random forest trees (NumTrees) is "300", the homogeneity criterion (Impurity) is "gini", the number of features to be considered for each node partition (FeatureSubsetStrategy) is "sqrt".

The software output is shown in Table 3 where the algorithm-based comparative results can be seen. The comparative evaluation of the results obtained from the used machine learning algorithms in terms of accuracy, training time, prediction time and difference between training-prediction time are as indicated below:

- **Accuracy and performance metrics:**
  Logistic Regression > Random Forest > Support Vector Machine > Naive Bayes
- **Training time:**
  Logistic Regression > Support Vector Machine > Random Forest > Naive Bayes
- **Prediction time:**
  Random Forest > Support Vector Machine > Logistic Regression > Naive Bayes
- **Difference between training-prediction time:**
  Logistic Regression > Support Vector Machine > Random Forest > Naive Bayes

Table 3. Machine learning algorithms performance evaluations

Algorithm	Accuracy	Precision	Recall	F-Measure	Training time (h)	Prediction time (h)
Logistic Regression	0.991	0.989	0.995	0.992	4.041	0.089
Support Vector Machine	0.939	0.928	0.939	0.933	1.419	0.092
Naive Bayes	0.809	0.914	0.809	0.858	0.016	0.026
Random Forest	0.980	0.972	0.985	0.978	0.412	0.297

If the accuracy and performance metrics are to be evaluated, it is seen that the highest correct prediction rate belongs to the Logistic Regression, Random Forest and Support Vector Machine algorithms. On the other hand, in the Logistic Regression algorithm, the training time and difference between training-prediction time has the highest values compared to other algorithms. This situation raises doubts as to whether the Logistic Regression algorithm is suitable for detecting attacks on the network.

When the obtained results in the study are compared with the results of previous intrusion detection system studies, it is concluded that Apache Spark has become increasingly more efficient with newer releases in detecting attacks on big network data.

In summary, the algorithms and the data used in similar studies in literature have been reviewed and their results were compared with the results of the Apache Spark tool used in this study. For future studies, the developed software can be carried out with different parameters through different machine learning algorithms with different network datasets. And comparative studies can be carried out with the results. Especially, with the developed software, intrusion detection systems using unsupervised classification algorithms that can detect newly encountered attacks can be developed. In terms of processing speed, machines with better properties will yield faster results. By analyzing the 41 features in the dataset, it is possible to increase the processing speed by removing data with less effect on classification. The study contributes to the literature in terms of examining and summarizing the performance of the new big data methods in the latest release of Apache Spark over network data.

6 Conclusion

The implemented study is an introduction to intrusion detection systems. The obtained results show that the Apache Spark tool has become increasingly effective in detecting attacks on big network data. This study is a guide for big data researchers and provide a reference source for the developed intrusion detection systems by comparing the performance of machine learning algorithms on network data.

As a continuation of this research, the next step would be to extract the features that have less effect on the classification in the dataset, followed by performance comparisons between the performed and the previously performed classification.

References

1. Çevik, M.: Intrusion detection with pattern classification. Ph.D. thesis, Istanbul Technical University, Institute of Science and Technology (2005)
2. Becerikli, Y.: Advanced pattern recognition. Doctorate Lecture, Computer Engineering Departmant, Kocaeli University, Kocaeli, Turkey (2016)
3. Gupta, G.P., Kulariya, M.: A framework for fast and efficient cyber security network intrusion detection using apache spark. Procedia Comput. Sci. **93**(Supplement C), 824–831 (2016)
4. Siddique, K., Akhtar, Z., Lee, H.G., Kim, W., Kim, Y.: Toward bulk synchronous parallel-based machine learning techniques for anomaly detection in high-speed big data networks. Symmetry **9**(9), 197 (2017)
5. Harifi, S., Byagowi, E., Khalilian, M.: Comparative study of apache spark MLlib clustering algorithms. In: Tan, Y., Takagi, H., Shi, Y. (eds.) DMBD 2017. LNCS, vol. 10387, pp. 61–73. Springer, Cham (2017). https://doi.org/10.1007/978-3-319-61845-6_7
6. Jeong, H.-D.J., et al.: A search for computationally efficient supervised learning algorithms of anomalous traffic. In: Barolli, L., Enokido, T. (eds.) IMIS 2017. AISC, vol. 612, pp. 590–600. Springer, Cham (2018). https://doi.org/10.1007/978-3-319-61542-4_58
7. Oh, S.W., Kim, H.S., Lee, H.S., Kim, S.J., Park, H., You, W.: Study on the multi-modal data preprocessing for knowledge-converged super brain. In: 2016 International Conference on Information and Communication Technology Convergence (ICTC), pp. 1088–1093. IEEE (2016)
8. Lightning-fast cluster computing. https://spark.apache.org/. Accessed 14 Mar 2018
9. Tavallaee, M., Bagheri, E., Lu, W., Ghorbani, A.A.: A detailed analysis of the KDD CUP 99 data set. In: 2009 IEEE Symposium on Computational Intelligence for Security and Defense Applications, CISDA 2009, pp. 1–6. IEEE (2009)
10. Intrusion Detector Learning. http://archive.ics.uci.edu/ml/machine-learning-databases/kddcup99-mld/task.html. Accessed 08 Jan 2018
11. Vapnik, V.: The Nature of Statistical Learning Theory. Springer, New York (2013). https://doi.org/10.1007/978-1-4757-3264-1
12. Özkan, Y.: Data Mining Methods. Papatya Publishing, Istanbul (2008)

13. Osuna, E., Freund, R., Girosi, F.: Support Vector Machines: Training and Applications. Massachusetts Institute of Technology, Cambridge (1997)
14. Pöyhönen, S.: Support vector machine based classification in condition monitoring of induction motors. Helsinki University of Technology (2004)
15. Ilhan Omurca, S.: Machine learning. Master Lecture, Computer Engineering Departmant, Kocaeli University, Kocaeli, Turkey (2016)
16. Akar, Ö., Güngör, O.: Classification of multispectral images using random forest algorithm. J. Geod. Geoinf. 1, 139–146 (2012)
17. Özdarıcı Ok, A., Akar, Ö., Güngör, O.: Classification of crops in agricultural lands using random forest classification method. In: TUFUAB 2011 VI. Technical Symposium, Antalya, Turkey (2011)
18. Gislason, P.O., Benediktsson, J.A., Sveinsson, J.R.: Random forests for land cover classification. Pattern Recogn. Lett. 27(4), 294–300 (2006)
19. Pal, M.: Random forest classifier for remote sensing classification. Int. J. Remote Sens. 26(1), 217–222 (2005)
20. Breiman, L.: Manual on setting up, using, and understanding random forests v3.1. Statistics Department, University of California Berkeley, CA, USA (2002)
21. Archer, K.J., Kimes, R.V.: Empirical characterization of random forest variable importance measures. Comput. Stat. Data Anal. 52(4), 2249–2260 (2008)
22. Breiman, L.: Random forests. Mach. Learn. 45(1), 5–32 (2001)
23. Hasan, M.A.M., Nasser, M., Pal, B., Ahmad, S.: Support vector machine and random forest modeling for intrusion detection system (IDS). J. Intell. Learn. Syst. Appl. 06, 45–52 (2014)

Predictive Models, Fuzzy and Recommender Systems

A Triangle Multi-level Item-Based Collaborative Filtering Method that Improves Recommendations

Gharbi Alshammari[1($\boxtimes$)], Stelios Kapetanakis[1,3], Nikolaos Polatidis[1], and Miltos Petridis[2]

[1] School of Computing, Engineering and Mathematics, University of Brighton, Moulsecoomb Campus, Lewes Road, Brighton BN2 4GJ, UK
{g.alshammari,s.kapetanakis,n.Polatidis}@brighton.ac.uk
[2] Department of Computer Science, Middlesex University London, The Burroughs, London NW4 4BT, UK
M.Petridis@mdx.ac.uk
[3] Gluru Research, 71-91 Aldwych, London WC2B 4HN, UK
stelios@gluru.co

Abstract. One of the most successful approaches that can provide a relevant recommendation in various domains is collaborative filtering. Although this approach has been widely applied, there are still limitations to be overcome in this research area. Accuracy is still one of the areas that need to be improved. In addition, the rapid growth of information available online presents recommender systems with several challenges. More specifically, data sparsity and coverage affect the quality of the recommendations that can be provided. In this paper, we propose an item-based collaborative filtering (IBCF) approach with triangle similarity measures that take into account the length and angle of rating vectors between users and allow positive and negative adjustments using a multi-level recommendation approach. We have improved the predictive accuracy and effectiveness of the proposed method, which outperforms all the compared methods in terms of the mean absolute error (MAE) and the root mean squared error (RMSE). We aimed to evaluate the proposed method by comparing our results with those of some popular similarity measures using k-nearest neighbour (kNN) algorithms. We ran our experiment using three real dataset: MovieLens 100K, MovieLens 1M and Yahoo! Movies.

Keywords: Collaborative filtering · Recommender systems
Triangle · Multi-level · Item-based

1 Introduction

Recommender systems help users and relevant items that meet their interests to address the problem of information overload. Moreover, users have trouble

© Springer Nature Switzerland AG 2018
E. Pimenidis and C. Jayne (Eds.): EANN 2018, CCIS 893, pp. 145–157, 2018.
https://doi.org/10.1007/978-3-319-98204-5_12

handling large volumes of information, and problems with cognitive and data sparsity when attempting to find appropriate information at the right time [4]. These systems play an important role in the growth of online information by filtering and recommending relevant items. The knowledge discovery approach can help with making a personalised recommendation by collecting a user's interests. Collaborative filtering (CF) is one of the most successful techniques for recommender systems [17].

Many collaborative filtering techniques have been proposed in different domains, such as e-commerce applications. Typically, elaborate approaches outperform the commonly-used k-nearest neighbour (kNN) baseline method in terms of accuracy, particularly for sparse datasets or in terms of scalability as they rely on offline pre-processing or model-building phases [6].

Most CF approaches analyse user ratings to determine the similarity between users and items. The similarity measure is important for finding accurate results in recommender system. However, it is challenging to determine distance measures in these systems in order to find similarities between users. Collaborative filtering is the most common applied algorithm through the kNN approach [11]. The key issue in this technique is how to calculate the similarity between users or items by finding similar shared interest. It is significantly rely on the rating aspect, which allow users to assign a high or low rating to a certain item based on their preference or dislike for it [13]. Many similarity measures have been adopted in recommender systems such as Pearson's Correlation Coefficient (PCC) [17] and Cosine [20] to provide recommendation based on the absolute ratings between users. Hence, modified similarity measures are one of the most important challenges to improve the prediction accuracy in recommender systems.

In this paper, we propose a new method that utilises triangle similarity measures with a multi-level algorithm. We consider both the length and angle of the rating vectors between users, as well as the constraints that modify users similarity assigning these with different levels. The main contributions of this paper are as follows:

1. We proposed a new recommendation method that combines triangle similarity measures with a multi-level recommendation technique.
2. We ran extensive experiments to show its effectiveness based on three real datasets, conducting a comparison with a baseline and a state-of-the-art alternative.

2   Related Work

CF is the most popular technique for recommender systems. It has been widely implemented in different domains such as movies [15] and music [24] to generate recommendations. It is a method of information filtering that seeks to predict the rating that a user will give to a particular item based on a similarity matrix. CF provided a foundation for the first recommenders systems, which were used to "help people make choices based on the opinions of other people" [7]. The task is to make an automatic prediction by considering other similar users ratings for an item. Therefore, the basic idea of CF is to find a user whose past rating behaviour is similar to that of the user the algorithm is currently trying to predict. This approach uses a kNN algorithm to calculate recommendations, and the main required data are the rating matrix and a function that computes similarity between users. An Item-based collaborative filtering (IBCF) technique looks into a set of items that the target users have rated and computes how similar they are to the target item; it then select the k most similar items. At the same time, the corresponding similarity are also computed [18]. Once the most similar items are found, the prediction is computed using a weighted average of the target users ratings. Hence, there are two main aspects to be considered: similarity computation and prediction generation [2]. Basically, to compute the similarity between items, the first step is to determine users who have rated both items and who have the most similar items with similar ratings. Many different measures can be used to compute the similarity between items, such as Pearson's correlation, cosine, Jaccard and Triangle similarity. Of these methods, Pearson's has had the most successful applications, which is defined in Eq. 1. Where $Sim_{a,b}$ is the similarity of users a and b, $r_{a,p}$ is the rating of user a for product p, $r_{b,p}$ is the rating of user b for product p and $\bar{r}a$, $\bar{r}b$ represent user's average ratings. P is the set of all products. However, In [21] the authors have proven that the triangle similarity outperforms Pearson correlation and shows improved results in IBCF.

$$\mathrm{Sim}_{a,b}^{PCC} = \frac{\sum_{P \in P}(r_{a,p} - \bar{r}a)(r_{b,p} - \bar{r}b)}{\sqrt{\sum_{P \in P}(r_{a,p} - \bar{r}a)^2}\sqrt{\sum_{P \in P}(r_{b,p} - \bar{r}b)^2}} \tag{1}$$

More recently, a combination of one or more methods called a hybrid recommender system has been applied to overcome the limitations of using one approach and obtain better results [5]. For instance, in [3], a hybrid case based reasoning approach was proposed to solve a long tail problem, which is items that have a few ratings by switching between collaborative filtering and content-based filtering. In addition, the authors in [12] implemented a hybrid recommender system that applied clustering technique and an artificial algae algorithm with a multi-level CF approach. However, co-rated items have been used for a problem solving in recommender systems to improve their predictive accuracy. Authors in [23] also introduced a hybrid approach for solving the problem of finding the rating of unrated items in a user-item matrix through a weighted combination of user-based and item-based collaborative filtering. These methods addressed the two major challenges of recommender systems, the accuracy of recommendations and sparsity of data, by simultaneously incorporating the correlation of users and items. In [22] the authors address a cold-start problem in user-based CF by considering both the distance between users and the co-rating of items using Jaccard factors. In [21], the authors proposed a new measure that integrates the triangle similarity approach with Jaccard similarity, which consider non co-rating users. The authors in [16] propose a multi-level constraint that improves the quality of a recommendation using PCC. Equation 2 considers the similarity between users relying on PCC and co-rated items in different levels.

$$
Sim_{a,b}^{PCC} = \begin{cases}
Sim_{a,b}^{PCC} + x_1, if \frac{|I_a \cap I_b|}{T} \geq t1 \ and \ Sim_{ij,iq}^{PCC} \geq y \\[2mm]
Sim_{a,b}^{PCC} + x_2, \ if \frac{|I_a \cap I_b|}{T} < t1 \ and \frac{|I_a \cap I_b|}{T}| \geq t2 \ and \ Sim_{a,b}^{PCC} \geq y \\[2mm]
Sim_{a,b}^{PCC} + x_3, \ if \frac{|I_a \cap I_b|}{T} < t2 \ and \frac{|I_a \cap I_b|}{T}| \geq t3 \ and \ Sim_{a,b}^{PCC} \geq y \\[2mm]
Sim_{a,b}^{PCC} + x_4, \ if \frac{|I_a \cap I_b|}{T} < t3 \ and \frac{|I_a \cap I_b|}{T}| \geq t4 \ and \ Sim_{a,b}^{PCC} \geq y \\[2mm]
0, \quad otherwise
\end{cases} \tag{2}
$$

3   The Proposed Method

The number of co-rated items reflects the degree of connection between users. For instance, a high number of co-rated items might indicate a high level of similarity. Traditional similarity metrics do not consider the number of co-rated items [19]. To solve this problem, a triangle similarity has been proposed by [21], which results in a significance improvement in accuracy when it is combined with co-rating. The triangle similarity is integrated with some constraints that apply a number of co-rated items.

In our approach, we apply a hybrid method that also adopts a multi-level CF approach, which enhances the similarity value of users that belong to certain categories and ignores the rest [16]. It enhances the process of kNN by finding a large margin within an application. The triangle similarity measure is defined as follows:

$$
Sim_{a,b}^{Tri} = 1 - \frac{\sqrt{\sum_{u \in c_{a,b}} (r_a - r_b)^2}}{\sqrt{\sum_{u \in c_{a,b}} r_a^2} + \sqrt{\sum_{u \in c_{a,b}} r_b^2}} \tag{3}
$$

The value range is [0, 1], where closer a value is to 1, the more similar they are. The triangle approach considers both the length of the vectors and the angle between them, so it is more reasonable than the angle-based cosine similarity.

For example if the two vectors $A = 5, 5, 5$ and $B = 1, 1, 1$ are given, then cosine similarity is 1. By contrast, the triangle similarity between them is 0.33 (Fig. 1).

$$
Sim_{a,b}^{Proposed} = \begin{cases}
Sim_{a,b}^{Tri} + x_1, if \frac{|I_a \cap I_b|}{T} \geq t1 \ and \ Sim_{ij,iq}^{Tri} \geq y \\[2mm]
Sim_{a,b}^{Tri} + x_2, \ if \frac{|I_a \cap I_b|}{T} < t1 \ and \frac{|I_a \cap I_b|}{T}| \geq t2 \ and \ Sim_{a,b}^{Tri} \geq y \\[2mm]
Sim_{a,b}^{Tri} + x_3, \ if \frac{|I_a \cap I_b|}{T} < t2 \ and \frac{|I_a \cap I_b|}{T}| \geq t3 \ and \ Sim_{a,b}^{Tri} \geq y \\[2mm]
Sim_{a,b}^{Tri} + x_4, \ if \frac{|I_a \cap I_b|}{T} < t3 \ and \frac{|I_a \cap I_b|}{T}| \geq t4 \ and \ Sim_{a,b}^{Tri} \geq y \\[2mm]
0, \quad otherwise
\end{cases} \tag{4}
$$

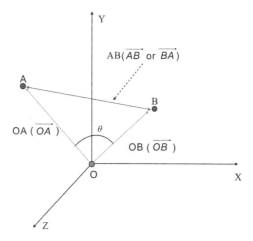

Fig. 1. A triangle in three-dimensional (3D) space [21].

In the above equation, $sim_{a,b}$ denotes the similarity between user a and user b. T stands for the total number of co-rated items. $t1$, $t2$, $t3$ and $t4$ are the predefined threshold of co-rated items for user similarity $Sim_{a,b}^{Tri}$. We consider that $t1 = 50$, $t2 = 20$, $t3 = 10$ and $t4 = 5$. We took $x1 = 0.5$, $x2 = 0.375$, $x3 = 0.25$, $x4 = 0.125$ and $y = 0.33$.

Compare the similarity given by $Sim_{a,b}^{Tri}$ using y and the number of co-rated items. If it is less than a specified level t, then go to the next level and continue go to the next level until the right level is found. If all four levels are not found, then the similarity is equal to 0.

4   Experimental Evaluation

Given below are the results for the three datasets with different parameters. All algorithms were implemented in the Java programming language. In this experiment, k represents the number of nearest neighbours.

4.1   Real Dataset

We have run our experiment with three real datasets in order to compare the results with different parameters, such as the number of items and users. All datasets have been evaluated using cross-validation with 5 folds and K is the number of neighbours, specified to be equal to 3, 10, 30, 50 and 100.

MovieLens 100K: This is a real dataset that is publicly available. It uses a web-based research recommender system that was conducted from September 1996 to April 1998. It contains 943 users and 1,682 movies. Each user has rated at least 20 movies. It contains 100,000 ratings, all of which are in a range between 1 and 5. The three main features are [UserID], [MovieID] and [Rating] [8].

MovieLens 1M: This dataset contains 1,000,209 anonymous ratings of approximately 3,900 movies made by 6040 users. The University of Minnesota created an online movie recommendation system, and its items are rated by users who joined MovieLens in 2000. All ratings are in a scale between 1 and 5. This dataset is also publicly available for running offline experiments and is widely used for collaborative filtering recommender systems [8].

Yahoo! Movies: This is a dataset obtained from Yahoo Labs under license. It contains 7,642 users, 11,915 movies and 211,111 ratings. The rating scale is between 1 and 5 [1].

4.2 Comparison

We ran the following IBCF algorithms to make a comparison between the following methods. All the methods used are detailed in the following sections:

PCC: In this method, the statistical correlation between the similar ratings of two users is calculated to find users that are the closest to a particular user. The output will be a value between −1 and 1; 1 is a totally positive correlation, 0 indicates that there is no correlation and −1 is a totally negative correlation.

Multi-level CF: This is a method that calculates the statistical correlation between the similar ratings of two users to find users that are the closest to a particular user. The output will be a value between −1 and 1; 1 is a totally positive correlation, 0 indicates that there is no correlation and −1 is a totally negative correlation.

The above two methods are compared in our proposed method. Our method considers both the length and angle of the rating vectors between users. Multi-level approach also considers the right level for each user after calculating the triangle similarity and compare it with a specified threshold. In addition, in each level, a certain constraint is conducted to modify the similarity between certain users who share similar items.

4.3 Evaluation Metrics

Recommender system researchers have applied different measures to evaluate the quality of proposed recommendation algorithms [10]. Since 1994 [17], most of the empirical studies examining recommender systems have focused on appraising the accuracy of these systems using different methods [9]. Appraisals of accuracy are useful for evaluating the quality of a system and its ability to forecast the rating for a particular item. Predictive accuracy measurement metrics are widely used by the research community in CF, which measures the similarity between true user ratings and recommender system predicted ratings. Hence, we apply both the mean absolute error (MAE) and the root mean squared error ($RMSE$) to measure the performance of the proposed methods and evaluate their prediction accuracy compared with other recommendation techniques. The MAE is defined in Eq. 5 and the RMSE is defined in Eq. 6.

$$\text{MAE} = \frac{1}{n} \sum_{i=1}^{n} |p_i - r_i| \tag{5}$$

$$\text{RMSE} = \sqrt{\frac{1}{n} \sum_{i=1}^{n} (p_i - r_i)^2} \tag{6}$$

In the above equations, p_i is the predicted rating, and r_i is the actual rating. It should be considered that lower values provide better result.

4.4   Experiment and Results

The MAE results in Fig. 2 show that when the number of neighbours is small, the prediction is significantly improved. For example, when $k = 3$ the PCC is $= 1.002$; in multi-level CF, is 0.83 whereas in our proposed method, it is $= 0.82$. In addition, our method outperforms other methods in all cases, even when the number of neighbourhood is high. In Fig. 3, we can see that the results of using the RMSE are significant in our method at $K = 30$ and 50.

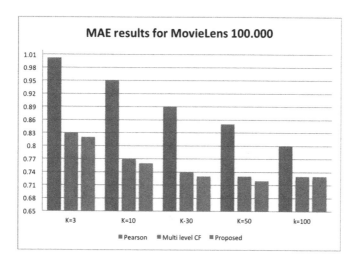

Fig. 2. MAE results for the MovieLens 100K dataset.

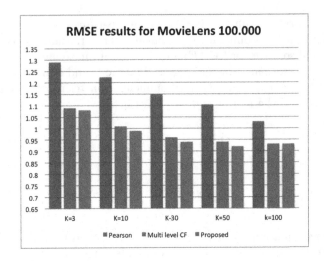

Fig. 3. RMSE results for the MovieLens 100K dataset.

Figures 4 and 5 show the MAE and RMSE results using MovieLens 1M ratings. In these figures, the improvement of our method compared with other methods is clear. However, when the number of neighbourhood is greater than 50, the multi-level CF value becomes close or equal to the result of our proposed method, such as $k = 100$. Overall, our proposed method achieves the minimal MAE in all cases.

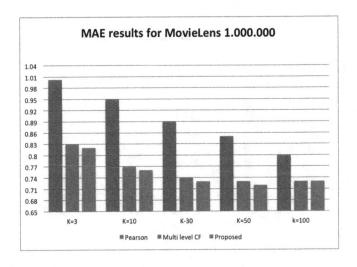

Fig. 4. MAE results for the MovieLens 1M dataset.

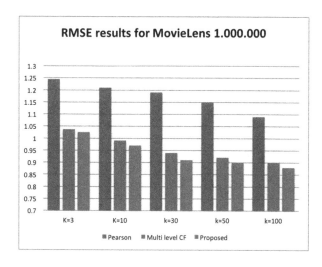

Fig. 5. RMSE results for the MovieLens 1M dataset.

For the Yahoo! Movies dataset shown in Figs. 6 and 7, we can see a significant change in all examined k value. For example, when $k = 3$, the baseline value is 0.85, the multi-level is 0.79 and the value in our method is 0.76. When $k = 100$ it can be seen that the difference between the three methods is slightly smaller but the proposed method still has the best results.

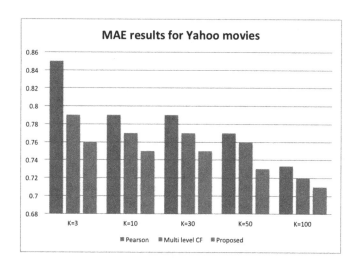

Fig. 6. MAE results for Yahoo! Movies dataset.

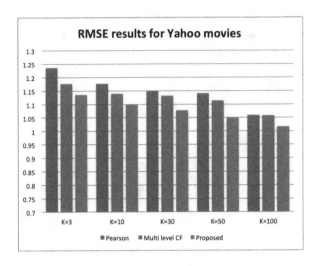

Fig. 7. RMSE results for Yahoo! Movies dataset

5 Discussion

In this paper we presented a new collaborative filtering method based on Triangle Similarity and multi-layer, multi-level similarity features. The proposed method has been experimentally evaluated using three real datasets and well known prediction accuracy metrics with the results being promising. In the MovieLens datasets it is shown that the proposed approach marginally outperforms the alternative, while in the Yahoo movies dataset the difference is higher between our method and the alternatives. Furthermore, the prediction error becomes smaller for each of the methods and in every dataset as the neighborhood is growing with the difference between our method and the alternative being similar in all cases. The proposed method outperforms the Pearson baseline and the multi-level CF state-of-the-art approach in all three datasets using both MAE and RMSE.

Although, the results are promising, they follow an "instance-base" recommendation approach, something that is promising in the short-term. However, they can pose severe limitations to richer, context-oriented user journeys. To overcome this limitation and as part of our future work we aim to use a Recurrent Neural Network (RNN) approach for pattern matching in collaborative filtering. RNNs are strong candidates for sequence matching and identification due to their consecutive structure and parsing of prev-current-next sequence entries for a given domain. RNNs are able to predict the "next" item in given several traces of the previous ones [14]. We propose an RNN approach compared to a possible probabilistic alternative (e.g. Hidden Markov Chains (HMC)) since it turn to minimise the number of iterations (only two data scans compared to multiple ones in the case of a HMC).

A single RNN will predict its expected output: y based on its hidden internal state: $h_0 .. h_{t-1}$ with an example can be seen in Fig. 8 below:

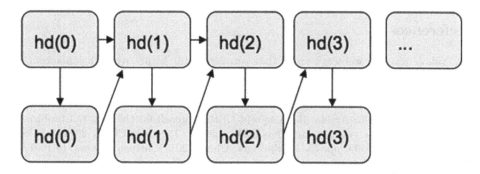

Fig. 8. RNN internal sequence prediction.

We assume sequences of (u_t, r_t) pairs where u_t is the user at time t and r_t is the desired recommendation. Such a pair is signal sequence and our dataset comprises several of them. For a recurrent network there is an additional input, a hidden state from the previous time step. We will represent it as h_{t-1}. A predicted output can be represented as r_t'. We define weight matrices as W_{hu} for the input to hidden layer weights, W_{hh} for hidden to hidden layer weights and W_{rh} for the hidden to output weights. From the above definitions We define a recurrent network as shown in Eqs. 7 and 8:

$$h_t = \sigma_h(W_{hh}h_t - 1 + W_{hu}u_t + b_h) \qquad (7)$$

$$r_t' = \sigma_r(W_{rh}h_t + b_r) \qquad (8)$$

Equations 7 and 8 can be applied to define a propagation function. To learn the weights we will generalise through time. Sequence prediction seems the first step towards a continuous time learning that should allow us to create the multiple levels dynamically, thus reducing the possibility of misbehavior.

6   Conclusions and Future Work

In this paper, we proposed a novel IBCF method based on triangle similarity and multiple similarity levels. The proposed method has been experimentally evaluated using three real datasets, along with a comparison to the traditional baseline PCC and to the state of the art multi-level CF. The results clearly show that the quality of the recommendations is improved in all cases and that our proposed method outperforms the alternatives.

Recommender systems have been widely used in different domains, including e-commerce and social media, so it is important to provide the best recommendations possible. Therefore, we have concentrated the best recommendations possible by reducing the prediction error, which results in high-quality recommendations. Furthermore, in the future, we aim to improve our method by making it dynamic and by performing a broader experimental evaluation.

References

1. Yahoo! research webscope movie data set. version1.0. http://research.yahoo.com/
2. Aggarwal, C.C., et al.: Recommender Systems. Springer, Cham (2016). https://doi.org/10.1007/978-3-319-29659-3
3. Alshammari, G., Jorro-Aragoneses, J.L., Kapetanakis, S., Petridis, M., Recio-García, J.A., Díaz-Agudo, B.: A hybrid CBR approach for the long tail problem in recommender systems. In: Aha, D.W., Lieber, J. (eds.) ICCBR 2017. LNCS (LNAI), vol. 10339, pp. 35–45. Springer, Cham (2017). https://doi.org/10.1007/978-3-319-61030-6_3
4. Bobadilla, J., Ortega, F., Hernando, A., Gutiérrez, A.: Recommender systems survey. Knowl. Based Syst. **46**, 109–132 (2013)
5. Burke, R.: Hybrid recommender systems: survey and experiments. User Model. User Adap. Inter. **12**(4), 331–370 (2002)
6. Gedikli, F., Jannach, D.: Recommending based on rating frequencies: accurate enough? In: Proceedings of the 8th Workshop on Intelligent Techniques for Web Personalization & Recommender Systems at UMAP10, pp. 65–70 (2010)
7. Goldberg, D., Nichols, D., Oki, B.M., Terry, D.: Using collaborative filtering to weave an information tapestry. Commun. ACM **35**(12), 61–70 (1992)
8. Harper, F.M., Konstan, J.A.: The movielens datasets: history and context. ACM Trans, Interact. Intell. Syst. (TiiS) **5**(4), 19 (2016)
9. Herlocker, J.L., Konstan, J.A., Borchers, A., Riedl, J.: An algorithmic framework for performing collaborative filtering. In: Proceedings of the 22nd Annual International ACM SIGIR Conference on Research and Development in Information Retrieval, pp. 230–237. ACM (1999)
10. Herlocker, J.L., Konstan, J.A., Terveen, L.G., Riedl, J.T.: Evaluating collaborative filtering recommender systems. ACM Trans. Inf. Syst. (TOIS) **22**(1), 5–53 (2004)
11. Jeong, B., Lee, J., Cho, H.: Improving memory-based collaborative filtering via similarity updating and prediction modulation. Inf. Sci. **180**(5), 602–612 (2010)
12. Katarya, R., Verma, O.P.: Effectual recommendations using artificial algae algorithm and fuzzy c-mean. Swarm Evol. Comput. **36**, 52–61 (2017)
13. Konstan, J.A., Riedl, J.: Recommender systems: from algorithms to user experience. User Model. User Adap. Inter. **22**(1–2), 101–123 (2012)
14. Mikolov, T., Karafiát, M., Burget, L., Černocký, J., Khudanpur, S.: Recurrent neural network based language model. In: Eleventh Annual Conference of the International Speech Communication Association (2010)
15. Miller, B.N., Albert, I., Lam, S.K., Konstan, J.A., Riedl, J.: Movielens unplugged: experiences with an occasionally connected recommender system. In: Proceedings of the 8th International Conference on Intelligent User Interfaces, pp. 263–266. ACM (2003)
16. Polatidis, N., Georgiadis, C.K.: A multi-level collaborative filtering method that improves recommendations. Expert Syst. Appl. **48**, 100–110 (2016)
17. Resnick, P., Iacovou, N., Suchak, M., Bergstrom, P., Riedl, J.: Grouplens: an open architecture for collaborative filtering of netnews. In: Proceedings of the 1994 ACM Conference on Computer Supported Cooperative Work, pp. 175–186. ACM (1994)

18. Sarwar, B., Karypis, G., Konstan, J., Riedl, J.: Item-based collaborative filtering recommendation algorithms. In: Proceedings of the 10th International Conference on World Wide Web, pp. 285–295. ACM (2001)
19. Shen, K., Liu, Y., Zhang, Z.: Modified similarity algorithm for collaborative filtering. In: Uden, L., Lu, W., Ting, I.-H. (eds.) KMO 2017. CCIS, vol. 731, pp. 378–385. Springer, Cham (2017). https://doi.org/10.1007/978-3-319-62698-7_31
20. Shi, Y., Larson, M., Hanjalic, A.: Collaborative filtering beyond the user-item matrix: a survey of the state of the art and future challenges. ACM Comput. Surv. (CSUR) **47**(1), 3 (2014)
21. Sun, S.B., et al.: Integrating triangle and jaccard similarities for recommendation. PloS One **12**(8), e0183570 (2017)
22. Tan, Z., He, L.: An efficient similarity measure for user-based collaborative filtering recommender systems inspired by the physical resonance principle. IEEE Access **5**, 27211–27228 (2017)
23. Wei, S., Zheng, X., Chen, D., Chen, C.: A hybrid approach for movie recommendation via tags and ratings. Electron. Commer. Res. Appl. **18**, 83–94 (2016)
24. Yoshii, K., Goto, M., Komatani, K., Ogata, T., Okuno, H.G.: An efficient hybrid music recommender system using an incrementally trainable probabilistic generative model. IEEE Trans. Audio Speech Lang. Process. **16**(2), 435–447 (2008)

Myo-To-Speech - Evolving Fuzzy-Neural Network Prediction of Speech Utterances from Myoelectric Signals

Mario Malcangi[1(✉)], Giovanni Felisati[2], Alberto Saibene[2], Enrico Alfonsi[3], Mauro Fresia[3], Roberto Maffiolettii[1], and Hao Quan[1]

[1] Computer Science Department, Università degli Studi di Milano, Milan, Italy
{malcangi,maffioletti,quan}@di.unimi.it
[2] Department of Heath Science, Università degli Studi di Milano, Milan, Italy
{giovanni.felisati,alberto.saibene}@unimi.it
[3] IRCCS Mondino Foundation, Pavia, Italy
{enrico.alfonsi,mauro.fresia}@mondino.it

Abstract. Voice rehabilitation is needed after several diseases, when a subject's vocal ability is compromised by surgical interference or removal of phonation organs (e.g. the larynx), by neural degeneration or by neurological injury to the motor component of the motor-speech system in the phonation area of the brain (e.g. dysarthria in Parkinson disease). A novel approach to voice rehabilitation consists of predicting the phonetic control sequence of the voice-production apparatus (larynx, tongue, etc.) by drawing inferences on the basis of myoelectric (EMG) signals captured by a set of contact electrodes, applied to the neck area of a subject with important phonatory alteration (e.g. laryngectomised) and intact neural control. The inference paradigm is based on an EFuNN (Evolving Fuzzy Neural Network) that has been trained to use the sampled EMG signal to predict the phoneme that corresponds to the motor control of the sublingual muscle movements monitored at phonation time. A phoneme-to-speech synthesizer generates audio output corresponding to the utterance the subject has tried to enunciate.

Keywords: EFuNN · Evolving Fuzzy Neural Network · Voice dysarthria
Voice rehabilitation · Myoelectric signal

1 Introduction

Voice rehabilitation may be required in several conditions or post-surgical settings. The worst case scenario involving the phonation production organ is total laryngectomy. Voice rehabilitation for such patient is a challenging task [1] because of the total removal of this fundamental organ. On the other hand, when compared to other diseases that involve the voice production (e.g. neurological diseases) laryngectomised subjects maintain the integrity and the control of the remaining phonation organs, so a rehabilitation strategy could be implemented as non-surgical (esophageal speech or laryngophone) or surgical (esophageal puncture with voice prosthesis insertion).

© Springer Nature Switzerland AG 2018
E. Pimenidis and C. Jayne (Eds.): EANN 2018, CCIS 893, pp. 158–168, 2018.
https://doi.org/10.1007/978-3-319-98204-5_13

Laryngectomy drastically alters the speech production capabilities of human beings, since for speech production three main physiologic elements are necessary: the power source (lung air), the sound source (larynx) and the sound modifier (vocal tract). The only element that is active after laryngectomy is the sound modifier that is controlled by the brain.

Three main options are available to restore voice after laryngectomy: laryngophone speech (non-surgical, apparatus-based), esophageal speech (non-surgical, apparatus-free), and tracheoesophageal speech (surgical, hand-held device-based).

The non-surgical approach is interesting because it could be fully under the control of the subject without requiring hospitalization. In the esophageal speech production the air is swallowed into the esophagus and then released, inducing a vibration of the pharyngeal mucosa. This vibrations are modulated and articulated by the tongue movements and the control of the oral cavities.

This rehabilitation method is completely noninvasive, but it is difficult to learn (only 20% of laryngectomised patients succeed in this endeavor).

The laryngophone (or electrolarynx) is a non-surgical device-based approach that try to mimic electronically the larynx functionality of subjects who lost the larynx after the surgical removing.

The electrolarynx is a vibrating devices that is applied to the submandibular region and induces on the air in the oral cavities the vibration at a frequency mimicking that produced by the vibrating folds of the larynx. This fundamental vibration is modulated and articulated by the tongue and other mouth muscles producing an intelligible utterance.

The use of this machine requires a training phase and the use of the hands that holds the device in contact with the neck (it is almost impossible to talk to over the telephone). Other disadvantages concern the voce quality: the voice quality is metallic and unnatural. Furthermore, it is not applicable if the skin is not sound conductive. To minimize the disadvantages of the traditional electrolarynx device as voice prosthesis some investigations occurred in the past using electro-myoelectric activity to synchronize the device with the brain control of vocal tract at utterance-time [2].

The electrolarynx (Fig. 1) is the real demonstration that the best approach to voice rehabilitation in laryngectomised subjects consists in the reuse of the preserved voice control ability (vocal tract motion control). Following this idea we assumed that the myoelectric control of the phonation organs is driven by the phonetic information of the speech utterance. In the brain it exists a mapping between the language phonemes and the language words set, so when a word is to be uttered the corresponding phones sequence is predicted in terms of myoelectric control of voice articulation muscles (mainly the tongue).

If for each phoneme of an uttered word a corresponding myoelectric pattern sequence exists, then we could therefore theoretically predict such phoneme from the myoelectric pattern and use an articulatory phonetic speech synthesizer to electronically generate high-quality voice in laryngectomized subjects. This voice prosthesis would be hand-free operating, being fully brain-controlled.

Fig. 1. The Electrolarynx is a non-surgical device that mimics the larynx function in laryngectomised subjects by inducing air vibration in the vocal tract.

2   Physiology of the Phonation

Vocal production in mammals and humans originates from a complex mechanism. In a nutshell, the main mechanical interaction is between the air emitted by the lungs and then modulated through the main respiratory and accessory muscles (diaphragm, ster-nocleidomastoideus, intercostal muscles) and the action of the vocal cords and laryngeal structures on the expiratory flow [3, 4].

In humans the action of the intrinsic laryngeal muscles is rather complex and focuses on the rotation movements of the arytenoid cartilages, on the movements of antero-posterior displacement of the cricoaritenoid structures and on the variation of tension on the vocal cords, the latter resembling elastic bands with high vibratory capacity, able to develop vibrational frequencies that exceed even 100 Hz (Fig. 2). In particular, the abductor muscles of the vocal cords are the posterior cricoarytenoideus muscles; the tensor muscles of the vocal cords are tyroarynenoideus, Vocal and, mainly cricothyroid muscles; the adductor muscles of the vocal cords are the lateral cricoarytoenoideus and the interarytoenoideus muscles [5].

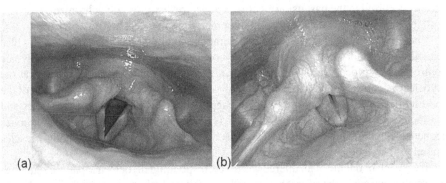

(a)   (b)

Fig. 2. Endoscopic view of an adult glottis, at rest (a) and during phonation (b), with medialization of the vocal folds in order to induce vibration and, consequently, sound formation.

The fundamental frequency and harmonics originated from the vocal folds vibration is then filtered inside the vocal tract (laryngeal cavity, pharynx, oral and nasal cavities) and modified by configurations and interactions of the articulators (e.g. tongue, lips and palate) into producing speech sounds which are naturally linked into forming words and sentences [6].

3 Neurology of the Phonation

The emotional vocalization of the mammals seems to originate from the circuit that reaches the central ponto-bulbar "central pattern generators" [7, 8] from the cortex of the cingulate gyrus and the peri-aqueductal gray substance. This vocalization can be modified by environmental conditions but it cannot be learned, being instinctive and, probably, linked to mechanisms that are related to the survival of the species. These types of vocalization originate from the phylogenetically older structures of the cerebral cortex (paleocortex), present at the level of the limbic lobe and are connected to subcortical structures. Conversely, the ability to speak, that is language, is learned; it is generative rather than imitative, since the human species is able to formulate new sentences for communication.

The language is integrated with the auditory system and with the systems of voluntary motor control [9, 10]. Only the human species has a direct cortico-bulbar pathway that originates from the laryngeal motor cortex and reaches the ambiguous nucleus [11]. A clinical model particularly useful for understanding the different mechanisms between communicative language and emotional vocalization is represented by a neurological pathology called spasmodic dysphonia (DSP). In the DSP the neural systems of learning of the voluntary spoken language are involved, while those of emotional vocalization are not. The left cortical peri-silvial system, connected to the ability to develop the spoken language, according to the structural and functional magnetic resonance studies, consists of the cortical areas represented by the supragarginal tour, the arched fasciculus, the frontal opercular area M1 and the internal capsule [12, 13]. Laryngeal muscles are bilaterally controlled by both hemispheres [14] thus making the system vulnerable to unilateral abnormalities that interfere with bilateral control of laryngeal muscles. Diffusion MRI techniques have shown that fractional anisotropy is reduced in the knee region of the internal capsule in SDRs and there is an increase in diffusivity bilaterally at the level of the cortico-bulbar tract [13]. Even basal ganglia regions such as putamen, globus pallidus, substantia nigra, the posterior arm of the inner capsule, but also the locus coeruleus, show degenerative neuropathological changes in patients with DSP. These patients have also described inflammatory changes in the structures of the reticulofalic reticular substance of the word "central pattern generators", although this region represents the final common path of emotional and voluntary vocalizations and is strange that is involved in a pathology such as DSP in only the voluntary component is involved. According to some authors it is hypothesizable that the truncated dysentery anomalies are related to the fact that the structures affected here are a selective part of circuits used for the correct control and propagation of the verbal timing in the spoken language and are not involved in emotional vocalizations [15]. If, in the patient with DSP, attempts are made to satisfy precise requests for control of the single words in the spoken language, anomalous compensatory mechanisms may also develop in the systems of production of the word at the cortical level [15].

4   Capturing and Processing the Myoelectric Signal

The myoelectric signal was detected by two surface electrodes (Ambu Neuroline 715 silver-cloridate 10 mm x 6 mm) set placed over the skin of the suprahyoid subchin muscles. These muscles represent the main complex for the elevation of laryngeal-pharyngeal structures during swallowing and speech, the neck sublingual muscles (infrahyoid muscles) that depress the hyoid bone and larynx during swallowing and speech. The electrodes distance was 30 mm (15 mm lateral to the middleline of the neck of each electrode). A third electrode (reference - ground) was placed on the shoulder (clavicle). The three-electrode set (Fig. 3) detects in differential mode the electric potential that controls the muscle during utterance of each phoneme of the word. The electrodes are connected to a computer-based electromyograph (Viasys Healthcare's Medelec Synergy SYN5-C) to display and record the signals applying the following setup:

- Channel 1: connected to electrodes
- Preprocessing: rectification, band pass filtering – 100–2000 Hz.

The recorded myoelectric patterns have been exported from the memory electromyographer and processed by a Matlab application (software program) to window the pattern corresponding to the uttered phoneme.

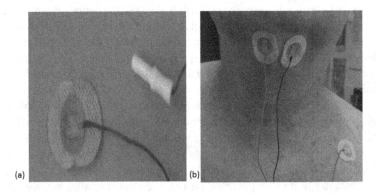

Fig. 3. Neurology surface electrode (Ambu Neuroline 715) (a) set placed over the skin of the suprahyoid subchin muscles and the reference electrode set placed over the clavicle (b).

5   Predicting from the Myoelectric Signal

The myoelectric sublingual muscle control signal embeds in its patterns the control sequence that moves the tongue to utter the word's phonemes. These patterns (Fig. 4) can be captured at speaking time and labeled with the corresponding phonemes to build up a labeled dataset to train a machine learning predicting paradigm.

Fig. 4. Myoelectric signal captured synchronously at the utterance-time of the vocal /a/.

The training dataset consisted of the raw sampled data from the myoelectric signal and one label that classify the pattern as corresponding to a specific phoneme. Thousands of patterns need to be collected and classified to proceed to supervised training of the predicting paradigm (learning). After learning the paradigm will be able to predict from a myoelectric pattern the phoneme that the subject is to utter.

To accomplish this task, two main systems need to be deployed:

- The data acquisition of the myoelectric signal
- The predicting paradigm.

The data acquisition of the myoelectric signal is a challenging task due to the very low voltage physical nature of such signals, the noninvasive requirements of this medical but non clinical application and the high presence of artifacts in the captured signal.

The predicting paradigm is also challenging because the requirements cannot be accomplished by the hardcomputing digital signal processing algorithms.

Traditional softcomputing methods (Neural networks and Fuzzy Logic) demonstrated to be effective and powerful in solving nonlinear pattern matching issues but not adequate for on-line, life-long learning through adaptation in a changing environment. The new framework named Evolving Connectionist Systems (ECOS) [16], specifically the Evolving Fuzzy Neural Network (EFuNN), capable to build intelligent agents, is adequate to execute the phoneme's prediction from the myoelectric signal at silent phonation time.

6   The Evolving Fuzzy Neural Network (EFuNN) Paradigm

The EFuNN [17, 18] (Fig. 5) is a particular implementation of the ECOS [16] (Evolving COnnectionist System) a biologically inspired framework [16]. It is a softcomputing

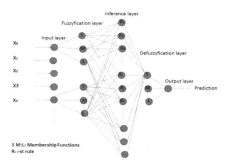

Fig. 5. EFuNN is a five layers Artificial Neural Network where each layer corresponds to a layer of a fuzzy logic engine.

paradigm that evolves through incremental, on-line learning, both supervised and unsupervised. EFuNN is order magnitude faster than multilayer perceptrons and fuzzy-neural networks.

EFuNNs are FuNNs that evolve according to the ECOS paradigm. FuNNs are Neural Networks that implements a set of fuzzy rules and the fuzzy inference engine in connectionist mode. FuNN is a feed-forward architecture-based with five layers of neurons and four layers of connections (Fig. 5). The first layer of neurons implements the input information layer. The second implements the membership layer that calculates the fuzzy membership degrees to which the input belongs to a fuzzy membership function. The third implements the fuzzy rules that encodes the associations between the input and the output data. The fourth calculates the degree to which the input data match the output membership functions. The fifth executes the defuzzification and calculates the crisp value for the output data. The FuNN is a combination of a Neural Network with a fuzzy engine The number of nodes and connections can change during the operation.

The peculiarity of EFuNN is its evolving capability and the one-pass learning. The nodes representing the membership function can be modified during the learning.

The rule nodes layer evolves through learning (supervised/unsupervised) that means all nodes are created/connected during learning. The nodes representing the membership functions can be modified at training-time. The same for the nodes representing the rules (input-output data association).

The evolving capability is incremental and adaptive. It is a bio-inspired way to make more effective the learning.

7   Dataset, Training and Test

To train and test the EFuNN's prediction capabilities related to the myoelectric patterns a dataset has been built. The dataset consists of several N-length raw sampled sequences of the myoelectric signal labeled by the corresponding phoneme code: $V_1 V_2 V_3 V_4 V_5 V_6 V_7 V_8 V_9 V_{10} V_{11} V_{...} V_j {}_{...} V_N L_n$

V_j: j-th amplitude of the j-th sample of the n-th myoelectric pattern
L_n: n-th label associated to the n-th sequence.

The N sequences composing the dataset are fed to the EFuNN settled in training mode. The EFuNN learns immediately from the data and it setups ready to be tested for prediction. A test dataset has been built in similar fashion of the training dataset. If the test is successful then the EFuNN is ready to run as phoneme predictor from myoelectric signal at silent speech production-time. If the test fails due to too many false predictions then the EFuNN is trained in evolving mode until the error rate is as low as required.

The first set of experiments to evaluate the EFuNN's phoneme prediction capabilities concerned the Italian language's vocal utterances (/a/ /e/ /i/ /o/ /u/ and the word /aiuola/).

The training dataset consists of myoelectric patterns captured and sampled synchronously with the utterance from an healthy subject.

The training consists only of the vocals, the test word is /aiuola/ that articulates almost all the vocals in a single word.

At training-time the EFuNN do not learn effectively. At test time it do not predict the right vocals sequence of the uttered word /aiuola/ under test (Fig. 6). After an evolving step (Fig. 7) the EFuNN learned to predict. At test-time it predicts without errors the right vocals phoneme sequence of the word /aiuola/: /a/ /i/ /u/ /o/ /a/ (Fig. 7).

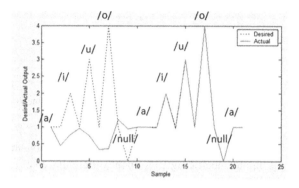

Fig. 6. Test /aiuola/ after one step training with single vocals /a/ /i/ /u/ /o/.

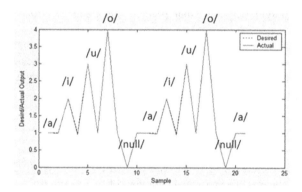

Fig. 7. Test /aiuola/ after one step evolving with single vocals /a/ /i/ /u/ /o/.

The modeling, training and test of the EFuNN have been executed with the simulation environment NeuCom developed at the Knowledge Engineering and Discovery Research Institute (KEDRI) Auckland – New Zeeland [19].

8   Myo-To-Speech System Framework

The Myo-to-Speech System Framework (Fig. 8) consists of three subsystems to transform the myoelectric signal from the sublingual muscle controlled by the phonation area of the subject's brain to the acoustical emission of the utterance.

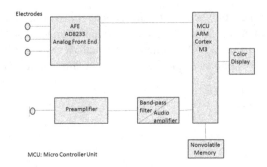

Fig. 8. Block chart of the prototyped myo-to-speech framework.

The first subsystem is the myoelectric signal acquisition AFE (Analog Front-End) consisting of the contact electrodes connected to an instrumentation amplifier and a set of filters to derive the bipolar electric signal from the differential electric signal captured by the contact electrodes.

The second subsystem is the Mixed-Signal MCU (Micro Controller Unit) that samples and collects in numeric format the myoelectric signal. This MCU executes also the inference paradigm (EFuNN) that predicts the phoneme from the myoelectric pattern.

The third subsystem is the phonemic speech synthesizer, that generates the electronic speech signal to be reproduced by a loudspeaker. The speech synthesizer is derived by the output of the predictor that encodes each predicted phoneme by a code embedded in the control part of the synthesizer.

9   Prototyping the Framework

To prototype the framework (Fig. 9) we used a set of fast prototyping COTS (Commercial Of The Shelf) boards. The most valuable part of the system is the analog front-end (AFE), that is available from Analog Devices in a CSP (Chip Scale Package) assembled on a prototyping COTS board (the AD8233 AFE). The AD8233 AFE is a fully integrated single lead electrocardiogram (ECG) analog front-end capable of 2–3 electrodes configuration with high signal gain (G = 100) with DC blocking capabilities and 80 dB (DC to 60 Hz) common rejection ratio. It integrates on a single-chip a 2-pole adjustable high-pass filter, one uncommitted operational amplifier and one 3-pole adjustable low-pass filter with adjustable gain.

A precision FET input unity-gain buffer (AD 8244) has been used to isolate source impedance from the of the signal chain.

Analog Devices AD8244's 2 pA maximum bias current, near zero current noise, and 10 TΩ input impedance introduce almost no error, even with source impedance well into the megaohms.

The AFE's input are connected to the three surface electrodes one reference and two differential. The AFE's output are connected to the bipolar input of the ADC (Analog to Digital Converter) of the MCU (MicroController Unit). The microcontroller unit is

Fig. 9. The prototype integrates all the function blocks of the system framework using COTS

the NXP K64F an ARM Cortex M4F MCU running at 120 MHz, an ultra-low power processor integrating a rich set of mixed-signal peripheral and a huge amount of fast and non-volatile memory available for data acquisition and processing. The M4 processing core enables the computing intensive processing requirements without increasing of power consumption. The MCU is integrated on the evaluation board NXP Freedom K64 F.

A microphone (preamplified, band –pass filtered and amplified) is also connected to a separate input of the MCU's ADC for synchronous acquisition of the utterance with the myoelectric signal.

An SD non-volatile memory is used to storage the sampled myoelectric signal and the uttered signal. A TFT touch screen display controlled by an FTDI VM800C multimedia controller is used for signal acquisition monitoring and human-machine interface (HMI).

10   Conclusion and Future Developments

Compared to other similar approaches [20], our method is innovative as it refers to latest microelectronic technologies, to the most promising inference paradigm (evolving) and because its invasivity has been minimized. Methods like that proposed in [20] are largely invasive because multiple surface electrodes are applied to subject's face, and no wearable electronics has been developed to minimize the degree of invasiveness.

The first round of research and developments leads to the deploying of a prototype device that enables the Myoelectric-To-Speech (MyoToSpeech) synthesis for voice rehabilitation in laryngectomized subjects. The tests demonstrated that the predicting paradigm is effective but several issues need to be solved, concerning the automatic segmentation and labeling of the myoelectric signal, the porting of the electronics to a wearable (patch) size, the building of the datasets for different languages, to preserve and synthesize the subject's original voice.

Acknowledgments. A special acknowledgment is due to Prof. Nikola Kasabov, Auckland University of Technology, Director KEDRI – Knowledge Engineering and Discovery Research Institute, for his invaluable suggestions on how to get the most from the EFuNN's evolving capabilities.

Acknowledgment is also due to Jan Hein Broeders (Analog Devices' healthcare business-development manager for EMEA) for his precious support and expertise in hardware prototyping, especially for the analog front-end (AFE) subsystem.

References

1. Balasubramanian, T.: Voice rehabilitation following total laryngectomy. Otoryngology Online J. **5**(1), 5 (2015)
2. Goldstain, E.A., Heaton, J.T., Kobler, B., Stanley, G.B., Hillman, R.E.: Design and implementation of a hand-free electrolarynx device controlled by neck strap muscle electromyographic activity. IEEE Trans. Biomed. Eng. **51**, 325–332 (2004)
3. Titze, I.R.: The physics of small-amplitude oscillation of the vocal folds. J. Acoust. Soc. Am. **83**, 1536–1552 (1988). https://doi.org/10.1121/1.395910. PMID 3372869
4. Lucero, J.C.: The minimum lung pressure to sustain vocal fold oscillation. J. Acoust. Soc. Am. **98**, 779–784 (1995)
5. Lucero, J.C.: Optimal glottal configuration for ease of phonation. J. Voice **12**, 151–158 (1998)
6. Mor, N., Simonyan, K., Blitzer, A.: Central voice production and pathophysiology of spasmodic dysphonia. Laryngoscope **128**(1), 177–183 (2018). Epub 23 May 2017, Review (2018)
7. Jürgens, U.: Neural pathways underlying vocal control. Neurosci. Biobehav. Rev. **26**, 235–258 (2002)
8. Jürgens, U.: A study of the central control of vocalization using the squirrel monkey. Med. Eng. Phys. **24**, 473–477 (2002)
9. Vihma, Mn., de Boysson-Bardies, B.: The nature and origins of ambient language influence on infant vocal production and early words. Phonetica **51**, 159–169 (1994)
10. Mac Neilage, P.F.: The frame/content theory of evolution of speech production. Behav. Brain Sci. **21**, 499–511 (1998)
11. Kuypers, H.G.: Cortico-bulbar connexions to the pons and lower brainstem in man: an anatomical study. Brain **81**, 364–388 (1958)
12. Haslinger, B., Erhard, P., Dresel, C., Castrop, F., Roettinger, M., Ceballos-Baumann, A.O.: Silent event-related, fMRI reveals reduced sensorimotor activation in laryngeal dystonia. Neurology **65**, 1562–1569 (2005)
13. Simonyan, K., Ludlow, C.L.: Abnormal activation of the primary somatosensory cortex in spasmodic dysphonia: an fMRI study. Cereb. Cortex **20**, 2749–2759 (2010)
14. Rödel, R.M., et al.: Human cortical motor representation of the larynx as assessed by transcranial magnetic stimulation (TMS). Laryngoscope **114**, 918–922 (2004)
15. Ludlow, C.L.: Spasmodic dysphonia: a laryngeal control disorder specific to speech. J. Neurosci. **31**(3), 793–797 (2011)
16. Kasabov, N.: Evolving Connectionist Systems: The Knowledge Engineering Approach. Springer, Heidelberg (2007). https://doi.org/10.1007/978-1-84628-347-5
17. Kasabov, N.: EFuNN, IEEE Tr SMC (2001)
18. Kasabov, N.: Evolving fuzzy neural networks – algorithms, applications and biological motivation. In: Yamakawa, T., Matsumoto, G. (eds.) Methodologies for the Conception, Design and Application of the Soft Computing, pp. 271–274. World Computing (1998)
19. http://www.kedri.aut.ac.nz/areas-of-expertise/data-mining-and-decision-support-systems/neucom
20. Zahner, M., Janke, M., Wand, M., Schultz, T.: Conversion from facial myoelectric signals to speech: a unit selection approach. In: Interspeech, pp. 1184–1188 (2014)

Model Prediction of Defects in Sheet Metal Forming Processes

Mario Dib[1]([✉]), Bernardete Ribeiro[1], and Pedro Prates[2]

[1] Center of Informatics and Systems of the University of Coimbra (CISUC),
University of Coimbra, Coimbra, Portugal
mario.dib@student.uc.pt, bribeiro@dei.uc.pt
[2] Center for Mechanical Engineering, Materials and Processes (CEMMPRE),
University of Coimbra, Coimbra, Portugal
pedro.prates@dem.uc.pt

Abstract. Predicting defects is a challenge in many processing steps during manufacturing because there is a great number of variables involved in the process. In this paper, we take a machine learning perspective to choose the best model for defects prediction of sheet metal forming processes. An empirical study is presented with the objective to choose the best machine learning algorithm that will be able to perform accurately this task. For building the model, three distinct datasets were created using numerical simulation for three mild steel materials: mild steel, DH600, HSLA340. The numerical simulation was performed on the basis of sixteen input features representing characteristics of the materials. Moreover, two kinds of defects, springback and maximum thinning, each one is binary with 1 (defects) and 0 (non-defects) were considered in the simulator. The experimental setup consists of running MLP, CART, NB, RF and SVM algorithms using cross-validation for correctly choosing model parameters. The results were averaged in 30 runs and the standard deviations recorded. The initial conclusion is that the learning algorithm scores differently depending on the type of defect and conditions of the experiment. Although the preliminary results show good performance of the algorithms in simulated environment, a further study with real data will be addressed in future work.

Keywords: Machine learning · Manufacturing process
Predictive model · Defect prediction · Algorithm comparison

1 Introduction

Sheet metal forming is a manufacturing process widely used in the production of metal components for the most diverse industries, namely the automotive, naval, aeronautical and machinery industries. However, sheet metal forming processes defects occur very often, making the overall procedure very costly; and the numerous variables involved in forming processes, related with the material

© Springer Nature Switzerland AG 2018
E. Pimenidis and C. Jayne (Eds.): EANN 2018, CCIS 893, pp. 169–180, 2018.
https://doi.org/10.1007/978-3-319-98204-5_14

properties, tooling geometry and process parameters, makes it difficult for engineers to accurately predict the occurrence of forming problems, such as cracking, localized necking, springback, among others. Also, the apparently random occurrence of forming defects due to sources of scatter (e.g. material properties, tooling geometry and process parameters) adds further complexity to the problem of defects prediction.

In this context, machine learning techniques may aid to solve this problem. In particular, they can be trained with available data for building defects prediction models. The rationale is that the models can generalize in unseen data and successfully identify and check such defect patterns.

Although it is certain that the machine learning technique is a viable method to be applied in the manufacturing industry to discover the origin of the defects, as described above, it is valid to point it out that there is, first, a necessity to discover how to build a model capable of learn the favorable conditions and actions for the defect to appear in a metal component.

Once this model is developed and with the knowledge of how to avoid the most common problems of the manufacturing of sheet metal, the process could be improved since the industry would be able to invest on the best materials for the production of certain pieces and with this simple measure it would also save time and money by not having to discard the defective pieces.

The focus on this paper is to create a model capable of learn about the conditions that will cause the defects to appear in a metal component, evaluate the results generated by the created model and perform a thorough comparison of machine learning models to verify which algorithms are capable to provide good results to solve the prediction problem. In summary, it aims to make a comparison of the results achieved by applying machine learning, through its algorithms, to predict defects in the manufacturing process of sheet metals.

In Sect. 2 the experimental sheet metal forming background will be described. The proposed approach used in this experiment will be detailed in Section 3. The results will be showed and discussed in Sect. 4. Finally, the conclusions are presented in Sect. 5.

2 Background

2.1 Sheet Metal Forming

Sheet metal forming is a manufacturing process that is widely used in the production of metal components for the most diverse industries, namely for the automotive, naval, aeronautical, machinery and household appliances industries [1]. In this process, metal sheets are plastically deformed into a desired shape by the action of forming tools, which typically consist of a punch, a die and a blank holder.

Firstly, a blank (i.e. non-deformed metal sheet) is placed over the die, and then it is pressed into the die by the movement of the punch to obtain the desired shape; the flow of the sheet material into the die is typically controlled with a blank holder. In general, sheet metal forming processes allow obtaining high quality components with high cadence and low cost; however, the variability inherent to mechanical properties, tool geometry and process parameters makes formed components often prone to defects such as wrinkling, tearing, excessive thinning and springback.

These defects can appear in any step of a forming process, and it can be very difficult to predict the location and the moment that they occur due to the large amount of variables involved in the process. The ever-increasing competitiveness in the automotive and aeronautical industries has demanded very high quality and robustness requirements, particularly in components that have a direct impact on occupant safety.

In this regard, the Finite Element Method (FEM) is a well-established computational tool that plays a key role in predicting defect-prone regions in components. Recently, some authors integrated statistical descriptions of variability within FEM, for assessing the sensitivity of defect predictions to scatter [1–3]. However, the integration of finite element analysis with statistical tools involves a high computational effort, incompatible with the response time required by the industry.

The design of the forming process of a metal component requires a high level of knowledge and is still very dependent on industrial experience. The use of computational tools to support design (CAD), planning (CAPP), engineering (CAE) and manufacturing (CAM) has contributed to facilitate analysis and reduce design time. However, it is always necessary for specialists to make decisions at different stages.

2.2   Machine Learning

In a changing environment, a system should be able to learn and adapt to such changes. In other words, the system needs to be intelligent. With the ability to learn, the system designer does not have to foresee the solutions for all possible situations [4]. In this context machine learning is a powerful tool to handle such problems.

To perform a thorough data analysis that automates the development of models, algorithms that learn interactively from example data or past experience are used, allowing computers to find patterns without being explicitly programmed [4]. These models may be predictive or descriptive, or both [4], and they should be able to learn how to predict defects based on previously data and independently adapt when exposed to a new data.

Machine learning techniques can be classified in two categories: supervised and unsupervised learning. In the first category, the algorithms use labeled data to train and learn, while in the unsupervised category they don't use data with a classification of observation [5]. In this paper we have only used supervised techniques.

There are a few studies in the literature on the application of machine learning for the identification and/or prediction of defects in metal manufacturing processes such as casting [6], rolling [7,8], extrusion [9] and sheet metal form-ing [10,11]. However, to our knowledge, no comparative study was performed so far on different machine learning algorithms for the prediction of defects in sheet metal forming processes.

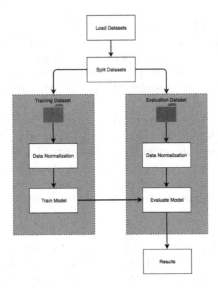

Fig. 1. Detection method scheme.

3   Model Defects Prediction Approach

3.1   Experimental Setup

The proposed scheme in Fig. 1 is composed by two sections, one for training the model and the other for model's evaluation. Both sections use the same dataset, split in training and evaluation data, and the same experiment configuration. Notice that 10-fold cross validation is used for setup of the model parameters. For this experiment, the following configuration was used:

- Dataset training size: 70%;
- Dataset evaluation size: 30%;
- Random seed: 7;
- Number of splits: 10.

The data normalization was performed after the configuration of the exper-iment because some algorithms would provide a better result with all the data at the same scale. The next step was the algorithms selection to be evaluated in regard of the best performance to identify the defect's occurrence patterns. Each selected algorithm was trained with the training data and later the models were evaluated with the new data provided by the validation data. The learned model uses the sample data and predicts whether a given sheet metal material belongs to the defect or non-defect class.

3.2   Experimental Datasets

Three distinct datasets were created based on numerical simulation results of the U-channel forming process, as schematized in Fig. 2, each dataset referring to one type of sheet material. The U-channel tooling comprises three main elements: the blank-holder, the die and the punch, where the blank-holder applies a constant force on the blank and the punch presses the metal sheet up to a total displacement of 30 mm, so that a plastic deformation occur in the blank to form a U-shaped part. The blank-holder force (BHF), which is the force that the blank-holder will apply on the blank, is assumed constant during the forming process and for this experiment the default values of 4.9 kN and 19.6 kN were used, leading to different responses.

To simulate this forming process, the BHF was applied on a combination of sixteen parameters of the forming process for each dataset, that are related with mechanical properties, in order to model variability in the mechanical properties of each material, which include: Young's modulus (E), anisotropy coefficients (r0, r45 and r90), initial tensile stress-strain data generated from parameters Y0, C and n (see Table 1) and initial sheet thickness (t0). The variables values were created using random numbers assuming that the variability in each input feature (excluding the blank holder force) is described by a normal distribution within a confidence interval of 95%, with well-defined mean and standard deviation values. The mean values for each input feature are those corresponding to the reference materials from Table 1; the standard deviation values were chosen in agreement with the literature cited in Prates et al. [3]. The effects' combination of the variables lead to springback and maximum thinning defects or to a result without any kind of defect.

Accordingly, three types of fictitious materials were studied, with mechanical behavior typical of mild, DP600 and HSLA340 steel sheets, as shown in Table 1 [12].

Table 1. Materials mechanical properties [12].

Variables	Mild steel	DP600	HSLA340
E [GPa]	206	210	210
v	0.3	0.3	0.3
r_0	1.790	1.010	0.820
r_{45}	1.510	0.760	1.070
r_{90}	2.270	0.980	1.040
Y_0 [MPa]	157.12	330.3	365.30
C [MPa]	565.32	1093.00	673
n	0.259	0.187	0.131
t_0 [mm]	0.78	0.78	0.78

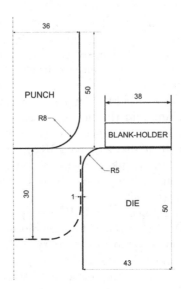

Fig. 2. U-channel scheme [3].

The two identified defects, springback and maximum thinning, were separated to create a binary classification with 1 (defects) and 0 (non-defects) to be considered in the simulator.

Accordingly, a defect exists if the final score is higher than the reference values shown in Table 2, which were obtained from the numerical simulations of the reference materials in Table 1. Both DP600 and HSLA340 had 170 FEM simulation runs performed, while Mild Steel had 174 FEM simulation runs performed. All the numerical simulations were carried out with the DD3IMP in-house FEM code, developed and optimized for simulating sheet metal forming processes [13]. The dataset for this experiment was created with the values of each input variable with it's respective final score, for each defect type in each material type.

Table 2. Non-defect reference values.

Material	Springback sb [mm]		Maximum thinning th [%]	
	BHF = 4.9 kN	BHF = 19.6 kN	BHF = 4.9 kN	BHF = 19.6 kN
Mild steel	6.165	2.601	2.82	9.62
DP600	11.151	8.522	2.07	5.83
HSLA340	8.747	2.70	5.115	7.66

3.3   Defects Classifier Selection

There is no easy way to know which algorithm will have the best performance to solve a proposed problem. Usually, it is difficult to understand the factors that affect the performance of a specific algorithm on a problem well enough to make the decisions the algorithm selection problem requires with confidence [14].

Five algorithms were randomly selected through machine learning disciplines. Three of them were classification algorithms, one is a regression algorithm and the last one is a statistical learning algorithm. The selected algorithms for the research design is presented in the following list:

– Multilayer Perceptron (MLP) [15];
– Random Forest (RF) [16];
– Decision Tree (CART) [17];
– Naive Bayes (NB) [18];
– Support Vector Machine (SVM) [19].

The models were built from the scratch using python v3.6.2 and related libraries, such as SciPy Ecosystem and Scikit-learn, based on this methodology [20]. Six machine learning models were created for two defects types in each metal type. The data was normalized for all the algorithms performances because the models results when it used normalized data were better, specially for the Multilayer Perceptron algorithm.

In addition, for all the models the same configuration with random weights was used in order to be possible to perform the model's results comparison. A brief description of each algorithm can be seen below.

Multilayer Perceptron (MLP): A Multilayer Perceptron or MLP is a feed-forward artificial model for neural networks. Its goal is to map sets of input data onto a set of appropriate outputs, using multiple layers of nodes in a directed graph, with each layer fully connected to the next one. Each node is a neuron, also called as processing element, except for the input nodes, with a nonlinear activation function [15]. The technique used by this model to train the network is called backpropagation, which consists in calculates the gradient of a loss function with respect to all the weights in the network, so that the gradient is fed to the optimization method which in turn uses it to update the weights, in an attempt to minimize the loss function.

Random Forest (RF): Random Forests are a combination of tree predictors such that each tree depends on the values of a random vector sampled independently and with the same distribution for all trees in the forest. They can be used for classification, regression and other tasks, that operate by constructing a multitude of decision trees at training time and outputting the class that is the mode of the classes or mean prediction, classification and regression respectively, of the individual trees [16].

Decision Tree (CART): Decision Tree algorithms were originally intended for classification which constructs a flowchart-like structure where each internal node denotes a test on an attribute, each branch corresponds to an outcome of the test, and each external node denotes a class prediction. At each node, the algorithm chooses the best attribute to partition the data into individual classes. When decision tree induction is used for attribute subset selection, a tree is constructed from the given data. All attributes that do not appear in the tree are assumed to be irrelevant. The set of attributes appearing in the tree form the reduced subset of attributes [17].

Naive Bayes (NB): The Naive Bayes classifiers are probabilistic classifiers based on applying Bayes' theorem, greatly simplify learning by assuming that features are independent and important given class [18].

Support Vector Machine (SVM): SVMs are supervised learning methods used for classification, regression and outliers detection, which constructs a hyperplane or set of hyperplanes in a high or infinite-dimensional space. In other words, given labeled training data (supervised learning), the algorithm outputs an optimal hyperplane which categorizes new examples [19].

4   Results and Discussion

In order to assess the effectiveness of the proposed algorithm comparison, several experiments were carried out in classification under covariate shift. In particular, the performance for predicting defects was evaluated, by averaging the results over 30 runs and the standard deviations recorded, with the metrics accuracy, precision, recall, f1-score and AUC (area under the curve).

All the information of the small dataset samples were stored in lists and later the mean and the standard deviation were calculated. To analyze the model's results in the sense of the defect prediction, different types of material were considered as mentioned above. It is important when making decisions to evaluate how good the generalization in unseen data is. As the classification is binary, the ROC Curve is an adequate metric since it gives all the possible operation points to choose from [21].

Were used 10-fold cross validation on the data to train the model and the models were trained in a Unix environment with 2.7 GHz processor and 16 Gb Ram. In the Table 3, the algorithm with the best performance is marked as red.

The experiments have revealed overall good results, although for some classifiers the performance has shown to be below average. However, in the most of the runs, the classification algorithms provided the best results, either for the MLP, the CART or the RF, for all types of metal. With exception of the maximum thinning defect in DP600 metal, the score was very close to the second best model, which was built with the MLP algorithm. The best algorithms for all the models are illustrated in the Table 3.

Table 3. Experiment results

Algorithms	Accuracy		Precision		Recall		F1-Score		AUC	
	Mean	std	Mean	std	Mean	std	Mean	std	Mean	std
Mild Steel - Springback Results										
MLP	81%	0.012	81%	0.011	81%	0.013	81%	0.013	80.67%	0.012
CART	88%	0.026	89%	0.012	88%	0.026	88%	0.026	87.69%	0.026
NB	70%	-	72%	-	70%	-	69%	-	70.09%	-
RF	85%	0.039	85%	0.037	85%	0.04	85%	0.04	84.71%	0.038
SVM	85%	-	86%	-	85%	-	85%	-	85.04%	-
Mild Steel - Maximum Thinning Results										
MLP	92%	0.0089	92%	0.0094	92%	0.005	92%	0.0054	91.01%	0.008
CART	86%	0.023	86%	0.024	86%	0.024	86%	0.024	84.91%	0.024
NB	89%	-	89%	-	89%	-	89%	-	87.68%	-
RF	89%	0.013	90%	0.013	90%	0.012	89%	0.009	88.31%	0.011
SVM	91%	-	91%	-	91%	-	90%	-	89.30%	-
DP600 - Springback Results										
MLP	91%	0.010	92%	0.010	91%	0.010	91%	0.010	91.90%	0.010
CART	87%	0.017	87%	0.018	87%	0.018	87%	0.018	86.54%	0.015
NB	84%	-	85%	-	84%	-	84%	-	85,11%	-
RF	86%	0.024	86%	0.028	85%	0.024	85%	0.024	84.76%	0.022
SVM	92%	-	92%	-	92%	-	92%	-	92.01%	-
DP600 - Maximum Thinning Results										
MLP	94%	0.010	95%	0.006	93%	0.010	93%	0.010	95.20%	0.008
CART	91%	0.009	92%	0.010	91%	0.010	91%	0.010	92.11%	0.010
NB	94%	-	94%	-	94%	-	94%	-	94.12%	-
RF	97%	0.015	97%	0.015	97%	0.015	97%	0.015	96.42%	0.015
SVM	92%	-	92%	-	92%	-	92%	-	90%	-
HLSA340 - Springback Results										
MLP	85%	0.014	86%	0.013	85%	0.014	85%	0.014	82.60%	0.016
CART	88%	0.015	89%	0.019	88%	0.015	88%	0.015	86.49%	0.016
NB	84%	-	85%	-	84%	-	84%	-	81.77%	-
RF	93%	0.017	93%	0.018	93%	0.017	93%	0.017	91.69%	0.018
SVM	80%	-	81%	-	80%	-	80%	-	76.77%	-
HSLA340 - Maximum Thinning Results										
MLP	91%	0.019	92%	0.016	91%	0.020	91%	0.020	91%	0.021
CART	94%	-	94%	-	94%	-	94%	-	93,98%	-
NB	86%	-	87%	-	86%	-	86%	-	85,88%	-
RF	94%	-	94%	-	94%	-	94%	-	93.98%	-
SVM	80%	-	82%	-	80%	-	80%	-	79.63%	-

The results were obtained by each model and were analyzed individually. The final results showed that the classification algorithms successfully predicted the defects in the considered materials. However, we noticed that there was not unanimity and the SVM algorithm was the best predicting the springback defect in the metal DP600, but even so, the MLP score was only 0.12% worse than the SVM score, meaning both models could provide comparable The Table 4 demonstrate the greatest outcomes for each defect-material combination.

Although, in this first moment, it was not possible to evaluate the accuracy of machine learning models with real data due to the unavailability of the data, the final outcomes showed there are different models that performed better for each type of material and its associate defects. Therefore, the next steps will demand models adjustments to take the best of each model. In this way, a more confident conclusion of which machine learning algorithm is apt to predict defects more accurately in real scenarios.

Table 4. Model results

Material types	Defect class	Model	Score
Mild steel	Springback	CART	87.69%
	Max. thinning	MLP	91.01%
DP600	Springback	SVM	92.01%
	Max. thinning	RF	96.42%
HSLA340	Springback	RF	91.69%
	Max. thinning	RF and CART	93.98%

5 Conclusion

Based on this experiment results, it is possible to have more than one option to build a machine learning model that is able to produce satisfactory outputs in regard of the defect prediction in a manufacturing environment. Whilst most of the scores had similar results independently of the material type or the defect class, an argument can be made in favor of the classification algorithms, because they had the best performance results for the accuracy and AUC parameters. In this sense, it would be a safe selection to use them as a standard choice to execute this type of prediction.

Although some algorithms did not perform well in some environments, as in case of a specific material-defect combination, it could have happened because of the small size of the training dataset, since the machine learning algorithms overall could learn better with larger samples of data, in accordance with each specific situation. That is why the usage of larger samples of data is one aspect that could be improved in the future to achieve better results.

Another aspect to be improved in the future is the personal configuration for each model. For this experiment the models were not refined in order to obtain the best possible result. In fact, the standard configuration, with some adjustments, to make the models outputs to be comparable was used.

Besides of the improvements suggested above, this experiment was helpful because it provides useful insights of which are the best algorithms to perform predictions. In this project's context this is a good start point for further investigations.

Acknowledgements. This work was funded by the Portuguese National Innovation Agency (ANI), for the support under the project SAFEFORMING - Sistema Inteligente de Preveno de Defeitos em Componentes Estampados a Frio, co-funded by FEDER, through the program Portugal-2020 (PT2020) and by POCI, with reference POCI-01-0247-FEDER-017762. Pedro Prates was supported by a grant for scientific research from the Portuguese Foundation for Science and Technology (ref. SFRH/BPD/101465/2014). All supports are gratefully acknowledged.

References

1. Huang, C., Radi, B., Hami, A.: Uncertainty analysis of deep drawing using surrogate model based probabilistic method. Int. J. Adv. Manuf. Technol. **86**, 9–12 (2016). https://doi.org/10.1007/s00170-016-8436-4
2. Wiebenga, J.H., Atzema, E.H., An, Y.G., Vegter, H., Boogaard, A.H.: Effect of material scatter on the plastic behavior and stretchability in sheet metal forming. J. Mater. Process. Technol. **214**(2), 238–252 (2014). https://doi.org/10.1016/j.jmatprotec.2013.08.008
3. Prates, P.A., Adaixo, A.S., Oliveira, M.C., et al.: Numerical study on the effect of mechanical properties variability in sheet metal forming processes. Int. J. Adv. Manuf. Technol., pp. 1–20 (2018). https://doi.org/10.1007/s00170-016-8436-4
4. Alpaydin, E.: Introduction to Machine Learning, 3rd edn. The MIT Press, Cambridge, Massachusetts, London (2016)
5. Aleem, S., Capretz, L.F., Ahmed, F.: Benchmarking machine learning techniques for software defect detection. Proc. Int. J. Softw. Eng. Appl. (IJSEA) **6**. Western University, London, Ontario, Canada (2015)
6. Santos, I., Nieves, J., Penya, Y.K., Bringas, P.G.: Optimising machine-learning-based fault prediction in foundry production. In: Proceedings of IWANN 2009: Distributed Computing, Artificial Intelligence, Bioinformatics, Soft Computing, and Ambient Assisted Living. Caligny, France and Bamberg, Germany (2009)
7. Lieber, D., Stolpe, M., Konrada, B., Deuse, J., Morik, K.: Quality prediction in interlinked manufacturing processes based on supervised & unsupervised machine learning. Procedia CIRP **7**, 193–198 (2013). https://doi.org/10.1016/j.procir.2013.05.033
8. Siyang, T., Xu, K.: An algorithm for surface defect identification of steel plates based on genetic algorithm and extreme learning machine. Metals - Open Access Metall. J. **7**, 311 (2017). https://doi.org/10.3390/met7080311
9. Barcellona, A.: Neural network techniques for metal forming design. In: Proceedings of the Thirtieth International MATADOR Conference. Palgrave, London (1993)

10. Wang, J., Wu, X., Thomson, P.F., Flitman, A.: A neural networks approach to investigating the geometrical influence on wrinkling in sheet metal forming. J. Mater. Process. Technol. **105**, 215–220 (2000). https://doi.org/10.1016/S0924-0136(00)00534-3

11. Wenjuan, L., Qiang, L., Feng, R., Zhiyong, L., Hongyang, Q.: Springback prediction for sheet metal forming based on GA-ANN technology. J. Mater. Process. Technol. **187–188**, 227–231 (2007). https://doi.org/10.1016/j.jmatprotec.2006.11.087

12. Bouvier, S., Teodosiu, C., Maier, C., Banu, M., Tabacaru, V.: Selection and identification of elastoplastic models for the materials used in the benchmarks. 18-Months Progress Report of the Digital Die Design Systems (3DS) (2001)

13. Menezes, L.F., Teodosiu, C.: Three-dimensional numerical simulation of the deep-drawing process using solid finite elements. J. Mater. Process. Technol. **97**, 100–106 (2000). https://doi.org/10.1016/S0924-0136(99)00345-3

14. Kotthoff, L., Gent, I., Miguel, I.: A preliminary evaluation of machine learning in algorithm selection for search problems. In: Proceedings of The Fourth International Symposium on Combinatorial Search. University of St. Andrews, Scotland, UK (2011)

15. Collobert, R., Bengio, S.: Links between perceptrons, MLPs and SVMs. In: Proceedings of the 21st International Conference on Machine Learning. Banff, Canada (2004)

16. Breiman, L.: Random Forests. Statistics Department, University of California Berkeley, CA, USA (2001)

17. Han, J., Kamber, M., Jian, P.: Data Mining Concepts and Techniques, 3rd edn. Morgan Kaufmann, San Francisco (2011)

18. Rish, I.: An Empirical Study of the Naive Bayes Classifier. IBM Research Division, Thomas J. Watson Research Center, Yorktown Heights, NY, USA (2001)

19. Cortes, C., Vapnik, V.: Support-vector networks. Mach. Learn. **20**, 273–297 (1995). https://doi.org/10.1007/BF00994018

20. Brownlee, J.: Machine Learning Mastery With Python: Understand Your Data, Create Accurate Models and Work Projects End-To-End, 1.4th edn. Jason Brownlee, Melbourne, Australia (2016)

21. Fawcett, T.: An introduction to ROC analysis. Pattern Recogn. Lett. **27**, 861–874 (2006)

Selecting Display Products for Furniture Stores Using Fuzzy Multi-criteria Decision Making Techniques

Özer Uygun[1], İlker Güven[2], Fuat Şimşir[2(⊠)], and Mehmet Emin Aydin[3]

[1] Sakarya University, 54187 Sakarya, Turkey
[2] Karabuk University, 78050 Karabuk, Turkey
fuatsimsir@karabuk.edu.tr
[3] University of West of England, Bristol, UK

Abstract. Efficient marketing in which the right products are supplied to the right consumer plays a crucial role for a profitable business in the age of highly accessible and competitive global market. This fact enforces producers to clearly identify and analyze the needs of consumers and to display their products respecting locality based on customers' needs. The position of the business is strengthened within the market and its competiveness increases by supplying and displaying the products suitable to regional consumers' preferences. In this study, an integral fuzzy multi criteria decision making technique is proposed for an effective decision making process to select the most suitable display products to the consumers' needs and preferences. The approach has been applied to identify the most suitable set of modular furniture products to be displayed at a local store that locates in Bursa city of Turkey. The approach uses Fuzzy DEMATEL method to work out the interrelations of chosen criteria, which are weighted with Fuzzy ANP and finally suggest a rank-based list of products with Fuzzy PROMETHEE. The results are verified with the expert view and found very useful.

Keywords: Fuzzy DEMATEL · Fuzzy ANP · Fuzzy PROMETHEE
Product selection · Multi criteria decision making techniques

1 Introduction

Modern marketing environments exhibit interesting behavioral relationships of producers and consumer, where enterprises that offer their products or services to consumers can affect shaping the habits of the consumers while the consumers' behavior affects promoting products. Obviously, the popularity of products among the consumers is one of the main factor in decision-making if a particular product will be kept produced or opted out of the market. Once a product is no longer requested by the consumer to some extent, companies stop produce it. For this reason, products should be attractive to the consumer. The fact that the market has various cultural, ethnic and moral constructs makes it difficult to present the right product to the right consumer.

© Springer Nature Switzerland AG 2018
E. Pimenidis and C. Jayne (Eds.): EANN 2018, CCIS 893, pp. 181–193, 2018.
https://doi.org/10.1007/978-3-319-98204-5_15

In such a situation where a precise and careful market analysis requires that the "right products" meet the "right consumers", the choice of the "right product" also has a critical prescription. In this respect, selecting the most suitable set of products to display among hundreds of alternative products is not a straightforward decision due to the fact that the complexity of this problem grows with increasing product variety.

The problem taken under consideration requires a firm decision by the decision makers with respect to many criteria, where each imposes constrictions upon the prospective solutions with positive and negative impact. In addition, inter-criteria effects may also bring potential overheads into the decision process. Meanwhile, the resources such as the size of shop floor, labor time are limited and need to be cost-effective. Therefore, such problems are considered for multi-criteria decision making for a firm decision to achieve cost-effective outcome.

Main purpose of this study is to explore how to utilize the market analytics in selecting the best set of display products, which offers the customers the best suiting supply of the product and propose an efficient approach for identifying the right product to the right customer by considering store location and customer expectations with many respects in order to increase the sales. In addition, we also aim to maximize usage of store area, catching target customer group and enhancing competitiveness.

In this study, market analysis is used to consider the above-mentioned problem as a decision problem and to propose an approach based on fuzzy multi-criteria decision making methods integrally used in identifying the best set of modular furniture products to exhibit in a specific department store. Every product group that company produced has been considered as an alternative and evaluated according to the total 22 sub-criteria including investors, regions, concepts, targeted customers, store areas and competitiveness as the main criteria. The fuzzy DEMATEL method was used to reveal the causal and effectual interrelationships of the criteria. This is followed by calculating the weight of each criterion by applying the fuzzy ANP method. The fuzzy PRO-METHEE method was finally applied using the final weights and the alternatives were evaluated and ranked.

2 Literature Review

Decision problems are rather complicated problems, which require considering a number of decision variables to be evaluated with respect to a number of criteria. There is a relatively rich literature in handled decision problems including multi-criteria decision-making approaches. Multi-criteria decision making techniques are utilized for considering many industrial and social problems to benefit of the analytical perspective gained in decision analysis. Although multi-criteria decision making techniques have very sound track record in solving various problems, there has been no study found on use of multi-criteria decision making approaches used within the scope of market analysis for their products.

The DEMATEL method is a frequently used method to reveal the relations of the criteria with each other. The ANP method is a method that considers the dependencies and feedback for both within the criterion and among the criterion. For this reason, it provides a more realistic approach to the problem of decision making [1]. When the literature is examined, DEMATEL and ANP methods are often used together to complete each other. Pamučar et al. [2] have explained the relationships in the data set by applying the DEMATEL method to the data set they used in their work. Uygun and Dede [3] used fuzzy multi-criteria decision making techniques to evaluate the performance of the green supply chain in their work. Relations between the performance evaluation criteria of the green supply chain in their work were revealed using DEMATEL and the fuzzy ANP method was applied considering these relations.

One of the other multi-criteria decision making techniques used in the literature is PROMETHEE. Vulević and Dragović [4] used the PROMETHEE method to rank sub-water basins in their work. Gül et al. [5] have addressed the problem of material selection using a fuzzy logic-based PROMETHEE method in their work. Bongo et al. [6] have developed an approach in which air traffic control officers use the DEMATEL-ANP and PROMETHEE II methods for workload stress.

ANP and PROMETHEE are used for many decision problems such as selection of ERP system for small and medium-sized enterprises, to evaluate and select light commercial vehicles for white goods services, to suggest new car-leasing system by comparing it with the existed one [7–9].

Efe et al. [10] examined the ergonomic product concept selection using heuristic fuzzy TOPSIS method. They pointed out that using heuristic fuzzy logic would give more accurate results because customers use linguistic expressions in their product preferences. In that study on mobile phone preferences, which products are preferred by consumers in order to guide the producers are presented by the heuristic fuzzy set theory.

3 Proposed Approach

In the study, an approach is presented to assess the selection of products to be placed in store by combined fuzzy multi-criteria decision making techniques according to the market demands of the producer company in a particular region. The general steps of the proposed approach are given as in the Fig. 1. Each method uses the obtained data from previous method.

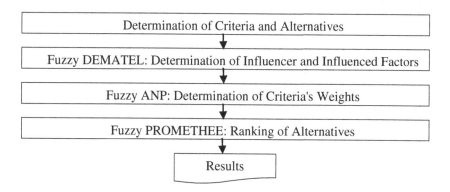

Fig. 1. Main steps of proposed approach for selecting display products in a Furniture Store

3.1   Fuzzy DEMATEL Method

The DEMATEL method is a model developed by the Geneva Battelle Institute as a method of revealing and analyzing the causality relationship between the factors in the model [11]. However, since it is difficult to quantify the interactions between the factors, the DEMATEL method has difficulty in determining the degree of relation between the factors. In order to come from this predicament, Lin and Wu have brought their problems fuzzy environment and presented the Fuzzy DEMATEL approach [12].

Step 1: Determination of criteria and creation of fuzzy scale. In the first step, the criteria to be applied should be determined. Two experts' views were taken into account in the determination of the criteria at the relevant firm as given in Fig. 2.

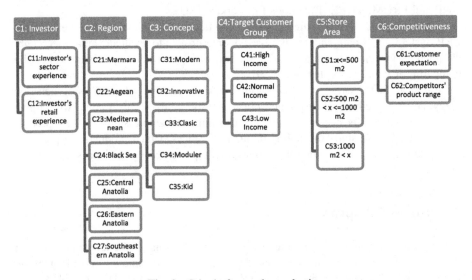

Fig. 2. Criteria for product selection

Step 2: Evaluating the criteria of the decision makers and establishing the direct relation matrix. In this step, n decision makers are asked to evaluate each criterion with the help of the linguistic expressions given in Table 1.

Table 1. The linguistic expressions and values used for the fuzzy DEMATEL method [13]

Linguistic expressions	Linguistic values
Very Low Influence (VLI)	(0,00; 0,00; 0,25)
Low Influence (LI)	(0,00; 0,25; 0,50)
Normal Influence (NI)	(0,25; 0,50; 0,75)
High Influence (HI)	(0,50; 0,75; 1,00)
Very High Influence (VHI)	(0,75; 1,00; 1,00)

As a result of this, n fuzzy evaluation matrices will be obtained. In order to proceed with the fuzzy DEMATEL process, the opinion of each expert is collected and divided into the number of experts and the fuzzy $\tilde{Z}$ matrix is obtained [3].

The initial direct relation fuzzy matrix contains fuzzy numbers. Here, each $\tilde{Z}_{ij} = \left(l_{ij}, m_{ij}, u_{ij}\right)$ is a triangular fuzzy number.

Step 3: Normalize the direct relation fuzzy matrix.

Step 4: After normalized direct relation fuzzy matrix, obtain the total relation fuzzy matrix.

Step 5: Each value of total relation fuzzy matrix in Step 4 are still triangular fuzzy numbers. Defuzzification must be done as directed in the literature to make them a single value.

3.2   Fuzzy ANP Method

The ANP method basically has the same structure as the AHP, but it also takes into account the relationships between the criteria. It was proposed by Saaty as an improved version of AHP. Unlike the AHP method, the ANP method aims to make an appropriate choice among the alternatives by evaluating the criteria in the horizontal plane with each other [3, 14–16].

As in the DEMATEL method, the ANP method is also expanded by the fuzzy set theory to remove the ambiguity in preferences. In the implementation phase of the fuzzy ANP method, the expansion analysis method proposed by Chang was followed as below [17].

Step 1: Calculate the synthetic expansion value S_i dependent on the ith goal.

Step 2: Calculate the preference between alternatives.

Step 3: Calculate the likelihood of a convex fuzzy number greater than k convex fuzzy numbers.

Step 4: Normalize the obtained values.

In this study, the Fuzzy ANP method is used to calculate the weights of the criterion according to the results of the total relation matrix obtained from the Fuzzy DEMATEL method. For the ANP, the linguistic expressions and values given in Table 2 were used during the taking of expert opinions.

Table 2. The linguistic expressions and values used in the fuzzy ANP method

Linguistic expressions	Linguistic values
Equally Important (EI)	(1, 1, 1)
Weakly Important (WI)	(2/3, 1, 3/2)
Strongly Important (SI)	(3/2, 2, 5/2)
Very Important (VI)	(5/2, 3, 7/2)
Absolutely Important (AI)	(7/2, 4, 9/2)

3.3   Fuzzy PROMETHEE Method

The combination of the PROMETHEE method with fuzzy logic was performed by Le Ténoa and Mareschal [18]. There is no change in terms of implementation between PROMETHEE and Fuzzy PROMETHEE methods. The actual difference in the fuzzy PROMETHEE method is the calculations to be done with fuzzy numbers.

When the fuzzy PROMETHEE calculations are performed, the LR type fuzzy numbers used in Yılmaz and Dağdeviren [19] studies, will be used and the Yager (1981) index will be taken into consideration [13]. Here $\tilde{F} = (n, a, b)$ is expressed as LR type fuzzy number. All numbers between $(n - a)$ and $(n + b)$ belong to the fuzzy cluster where a (L) and b (R) give the right and left spreading function of the n.

Step 1: In this study, the criterion and the weights of them will be obtained from Fuzzy DEMATEL and Fuzzy ANP methods and will form the data matrix with the alternatives. $w = (w1, w2 \ldots wn)$ express weights, $c = (f1, f2, \ldots, fn)$ express evaluations based on the criteria and $A = (a, b, \ldots, n)$ express alternative as shown in Table 3. Here, each $Fi(n)$ value is a LR type fuzzy triangular number.

Table 3. PROMETHEE initial decision matrix

Criteria	a	b	...	n	w
f1	$F_1(a)$	$F_1(b)$	...	$F_1(n)$	w1
f2	$F_2(a)$	$F_2(b)$	...	$F_2(b)$	w2
...	...	...	...	...	...
fn	$F_n(a)$	$F_n(b)$	...	$F_n(n)$	wn

Step 2: There are six types of preferred functions for PROMETHEE. It can be used for implementation by choosing the preference function that best suits the problem structure and best explains the problem. The linear type of preference function will be used in this study as in Eq. 1.

$$P(x) = \begin{cases} 0 & n - c \leq q \\ \frac{\tilde{x}(n,a,b)-q}{p-q} & q \leq n - c \, ven + d \leq p \\ 1 & n + d \geq p \end{cases} \tag{1}$$

Step 3: Based on the preference functions, the common preference functions for alternative pairs can be determined.

Step 4: The obtained matrix consists of fuzzy triangular numbers of the LR type. To defuzzification of such numbers, the Yager index is used.

Step 5: The preference indices of alternatives a and b is calculated. k criteria evaluates a and b alternatives with weights w_i (i = 1, 2, ..., k)

Step 6: For alternative a, the positive and negative advantages are calculated.

Step 7: Partial priorities specify the position of the alternatives against each other. Here, some alternatives can be compared with each other, while others are not.

Step 8: The PROMETHEE II method calculates the full priority values and allows the alternatives to be evaluated in the same plane by Eq. 2. Thus alternatives can be shown in a complete sequence from best to worse.

$$\Phi(a) = \Phi^+(a) - \Phi^-(a) \tag{2}$$

Linguistic expressions and values for PROMETHEE are given in Table 4.

Table 4. The linguistic expressions and values used in the fuzzy PROMETHE method

Linguistic variables	Linguistic values	Linguistic values arranged according to Yager Index
Very Poor (VP)	(0, 0, 0.15)	(0, 0, 0)
Poor (P)	(0, 0.15, 0.3)	(0.15, 0.15, 0.15)
Medium Poor (MP)	(0.15, 0.3, 0.5)	(0.3, 0.15, 0.2)
Fair (F)	(0.3, 0.5, 0.65)	(0.5, 0.2, 0.15)
Medium Good (MG)	(0.5, 0.65, 0.8)	(0.65, 0.15, 0.15)
Good (G)	(0.65, 0.8, 1)	(0.8, 0.15, 0.2)
Very Good (VG)	(0.8, 1, 1)	(1, 0.2, 0)

4   Case Study and Implementation

The method proposed in the study was applied to the store opened in Bursa region by the manufacturer which produces modular furniture. Among the 49 alternative products, Fuzzy DEMATEL method was first applied to find out affecting and affected criteria, then Fuzzy ANP method was applied considering the criterion obtained, to determinate weights of criteria, and finally Fuzzy PROMETHEE method was applied to rank alternatives for determining the most appropriate product families for the regional market and consumer demands.

Firstly, in order to reveal the relationships among the alternatives, criteria were suggested and evaluated for a senior white collar who is dealing with franchisers and interior designers who are responsible for the product's location in the store. The linguistic expressions belonging to one of the experts are given in Table 5. The evaluations made by the two experts are given in Table 6 with the mean values.

The direct relation fuzzy matrix is normalized. Then, the total relation fuzzy matrix is obtained as in Table 7.

Defuzzified total relation matrix is obtained, and shown in Table 8. The threshold value is determined as 0.353 according to the expert opinions.

Table 5. The linguistic expressions of the evaluation of one of the experts for the criteria

Criteria	I	R	C	TCG	SA	CM
I	0	VHI	LI	VLI	VHI	VLI
R	NI	0	VHI	HI	VLI	NI
C	LI	VHI	0	NI	NI	HI
TCG	HI	HI	VHI	0	VHI	NI
SA	VHI	LI	VHI	LI	0	NI
CM	NI	HI	VHI	VLI	VLI	0

Legend: Investor (I), Region (R), Concept (C), Target Customer Group (TCG), Store Area (SA), Competitiveness (CM)

Table 6. Both experts' evaluations are the resulting direct relation fuzzy matrix (some parts)

Criteria	I			R			C		
I	0	0	0	0,375	0,625	0,875	0	0,125	0,375
R	0,375	0,625	0,875	0	0	0	0,75	1	1
C	0	0,25	0,5	0,75	1	1	0	0	0

Table 7. Total relation fuzzy matrix (some parts)

Criteria	I			R			C		
I	0,0342	0,127	0,516	0,105	0,266	0,751	0,046	0,19	0,645
R	0,1094	0,294	0,846	0,077	0,255	0,81	0,219	0,434	0,945
C	0,0426	0,236	0,76	0,221	0,44	0,957	0,091	0,271	0,743

Table 8. Defuzzified total relation matrix

Criteria	I	R	C	TCG	SA	CM
I	0,201	0,347	0,268	0,191	**0,353**	0,187
R	**0,386**	0,349	**0,508**	0,344	0,309	**0,373**
C	0,319	**0,515**	0,344	**0,356**	0,298	**0,410**
TCG	**0,426**	**0,538**	**0,532**	0,253	**0,437**	**0,355**
SA	**0,417**	**0,425**	**0,469**	0,288	0,244	0,313
CM	0,29	**0,423**	**0,448**	0,208	0,251	0,215

After obtaining cause and effect relationship by the Fuzzy DEMATEL method, pairwise comparisons were made to calculate the criteria's weights by applying the Fuzzy ANP method. For example, the Investor criterion, as seen in Table 8, is affecting the Store Area criterion. The sub-criteria of the store area C51, C52 and C53 are evaluated by taking into account the sub criteria of the investor. In Table 9, one of the experts' evaluations is given for the store area according to the Investor Sector Experience sub criterion.

Table 9. Taking into consideration the investor sector experience criterion, the linguistic expression of the expert evaluation for the store area

Criteria	(C51)	(C52)	(C53)
(C51)	EI		
(C52)	EI	EI	
(C53)	SI	SI	EI

Table 10. Taking into consideration the investor sector experience criterion, the fuzzy values of the expert evaluation for the store area

Criteria	(C51)			(C52)			(C53)		
(C51)	1	1	1	1	1	1	0,4	0,5	0,667
(C52)	1	1	1	1	1	1	0,4	0,5	0,667
(C53)	1,5	2	2,5	1,5	2	2,5	1	1	1

Table 11. Considering the investor sector experience criterion, the geometric mean and calculated weights of the expert assessments made for the store area

Criteria	(C51)			(C52)			(C53)			Weights (W_i)
(C51)	1	1	1	0,816	1	1,225	0,338	0,408	0,516	**0,005**
(C52)	0,8165	1	1,225	1	1	1	0,775	1	1,291	**0,278**
(C53)	1,9365	2,449	2,958	0,775	1	1,291	1	1	1	**0,717**

Table 12. Unweighted super matrix

	C11	C12	C21	C22	...	C51	C52	C53	C61	C62
C11	0	0	0,924	0,672	...	0,5	0,672	0,672	0	0
C12	0	0	0,076	0,328	...	0,5	0,328	0,328	0	0
C21	0	0	0	0	...	0,083	0,105	0,091	0,23	0,163
⋮	...	...	...	...	...	...	...	...	...	...
C53	0,717	0,377	0	0	...	0	0	0	0	0
C61	0	0	0,662	0,924	...	0	0	0	0	0
C62	0	0	0,338	0,076	...	0	0	0	0	0

In Table 10, fuzzy number equivalents are given for this evaluation.

For each criterion above the threshold value, the sub-criteria were evaluated by the experts, and then the local weights calculated by taking the geometric mean of the expert opinions. The results obtained are shown in Table 11. Similarly, all calculations were done for all criteria and sub-criteria that passed the threshold in the total relation matrix as a result of the Fuzzy DEMATEL method.

The results obtained are given in the unweighted matrix as in Table 12, and limit super matrix is given as in Table 13.

Table 13. Limit super matrix

	C11	C12	C21	C22	...	C51	C52	C53	C61	C62
C11	0,097	0,097	0,097	0,097	...	0,097	0,097	0,097	0,097	0,097
C12	0,047	0,047	0,047	0,047	...	0,047	0,047	0,047	0,047	0,047
C21	0,038	0,038	0,038	0,038	...	0,038	0,038	0,038	0,038	0,038
:	...	...	...	...	...	...	...	...	...	...
C53	0,095	0,095	0,095	0,095	...	0,095	0,095	0,095	0,095	0,095
C61	0,098	0,098	0,098	0,098	...	0,098	0,098	0,098	0,098	0,098
C62	0,069	0,069	0,069	0,069	...	0,069	0,069	0,069	0,069	0,069

After calculating the weights of the criteria using the fuzzy ANP method, the alternatives were evaluated to apply the Fuzzy PROMETHEE method according to the criteria using the linguistic expressions given in Sect. 3.3. Table 14 lists the linguistic values of one of the experts for the alternatives. Similarly, after taking other expert opinions, a fuzzy initial decision matrix for Fuzzy PROMETHEE was generated by calculating the average of expert opinions. Pairwise comparisons were made for each alternative using the preference function given in Eq. 1, then common preference function is used, and the resulting matrix of common preference functions was refined using the Yager index. Following steps applied from Sect. 3.3 and finally, full priority values are calculated using Eq. 2 and the alternatives are listed as in Table 15. According to this table, among the all alternatives, the alternatives suitable for the store to be opened in Bursa province are Product Family (PF) 30, PF 48, PF 26, PF 18 and so on.

Table 14. Linguistic values of first five product family for Investor criteria

Alternatives	Investor (C1)					
	C11			C12		
PF.001	0,5	0,2	0,15	0,65	0,15	0,15
PF.002	0,65	0,15	0,15	0,65	0,15	0,15
PF.003	0,65	0,15	0,15	0,65	0,15	0,15
PF.004	0,8	0,15	0,2	0,65	0,15	0,15
PF.005	0,15	0,15	0,15	0,5	0,2	0,15

Legend: Product Family (PF)

Table 16 shows sales of two stores in the same area. Store 1's product selection was done according to given approach in this paper and Store 2's products were decision of the responsible person. Store 1's selected products cover %66 sales of total sales of the store where Store 2's product sales are only %31 of total sales of the store. It can be seen that catching customer with the right products can increase the sales.

Table 15. List of the selected product families by result of Fuzzy PROMETHEE

Rank	Alternatives	Φ+	Φ−	ΦNET
1	PF.030	0,186	0,018	**0,1679**
2	PF.048	0,182	0,022	**0,1600**
3	PF.026	0,192	0,035	**0,1570**
:	...	...	...	...
48	PF.020	0,012	0,301	**−0,2893**
49	PF.014	0,006	0,348	**−0,3421**

Table 16. List of the selected product families sales

Rank	Alternatives	Store 1 Sales (pcs)	Store 2 Sales (pcs)
1	PF.030	19	Non-exhibited
2	PF.048	11	16
3	PF.026	12	Non-exhibited
4	PF.018	8	9
5	PF.044	45	32
6	PF.035	38	40
7	PF.033	55	24
8	PF.038	27	Non-exhibited
9	PF.016	83	Non-exhibited
10	PF.037	19	Non-exhibited
Total sales of selected PF		**317**	**121**
Total sales of store		479	387
Total sales of selected PF over total sales (%)		%66,18	%31,27

5 Conclusion

The practice and research prove that the use of multi-criteria decision making techniques helps making firm decision to produce more beneficial outputs as each technique in this regard imposes use of analytical and effective comparative processes. Furthermore, embracing fuzzy logic to empower these techniques even further helps generate more realistic solutions with high reliability as it brings human experts' view in-the-loop for evaluations and judgments. In this study, an approach using a framework of multiple fuzzy multi-criteria decision making techniques is implemented for solving display products selection problems. The approach is applied to a real problem case that suggests a set of products to exhibit in a department store recently opened in Bursa province Turkey, which sells modular home and office furniture. The proposed approach has been successfully applied and the results were verified with human expert view; then a substantial list of display products are suggested to the company for other stores.

References

1. Ömürbek, N., Şimşek, A.: Analitik Hiyerarşi Süreci ve Analitik Ağ Süreci Yöntemleriyle Online Alışveriş Site Seçimi. J. Manag. Econ. Res. **22**, 306 (2014)
2. Pamučar, D., Mihajlović, M., Obradović, R., Atanasković, P.: Novel approach to group multi-criteria decision making based on interval rough numbers: Hybrid DEMATEL-ANP-MAIRCA model. Expert Syst. Appl. **88**, 58–80 (2017)
3. Uygun, Ö., Dede, A.: Performance evaluation of green supply chain management using integrated fuzzy multi-criteria decision making techniques. Comput. Ind. Eng. **102**, 502–511 (2016)
4. Vulević, T., Dragović, N.: Multi-criteria decision analysis for sub-watersheds ranking via the PROMETHEE method. Int. Soil Water Conserv. Res. **5**(1), 50–55 (2017)
5. Gul, M., Celik, E., Gumus, A.T., Guneri, A.F.: A fuzzy logic based PROMETHEE method for material selection problems. Beni-Suef Univ. J. Basic Appl. Sci. **7**(1), 68–79 (2017)
6. Bongo, M.F., Alimpangog, K.M.S., Loar, J.F., Montefalcon, J.A., Ocampo, L.A.: An application of DEMATEL-ANP and PROMETHEE II approach for air traffic controllers' workload stress problem: a case of Mactan Civil Aviation Authority of the Philippines. J. Air Transp. Manag. **68**, 198–213 (2017)
7. Kilic, H.S., Zaim, S., Delen, D.: Selecting 'The Best' ERP system for SMEs using a combination of ANP and PROMETHEE methods. Expert Syst. Appl. **42**(5), 2343–2352 (2015)
8. Ömürbek, N., Karaatlı, M., Eren, H., Şanlı, B.: AHP Temelli PROMETHEE Sıralama Yöntemi ile Hafif Ticari Araç Seçimi. Süleyman Demirel Üniversitesi İktisadi ve İdari Bilimler Fakültesi Dergisi, vol. 19, no. 4, pp. 47–64, December 2014
9. Öztoprak, E.: Kiralama Yoluyla Araba Temin Eden Bir İşletmede AHP Yöntemi Uygulaması. Atatürk Üniversitesi Sosyal Bilimler Enstitüsü Dergisi, vol. 18, no. 2, pp. 337–348, October 2014
10. Efe, B., Boran, F., Kurt, M.: Sezgisel Bulanık Topsis Yöntemi Kullanılarak Ergonomik Ürün Konsept Seçimi. Mühendislik Bilimleri ve Tasarım Dergisi, vol. 3, no. 3, pp. 433–440, December 2015
11. Karaoğlan, S.: DEMATEL Ve VIKOR Yöntemleriyle Dış Kaynak Seçimi: Otel İşletmesi Örneği. Akademik Bakış Uluslararası Hakemli Sosyal Bilimler Dergisi, no. 55, pp. 9–24, September 2016
12. Lin, C.-J., Wu, W.-W.: A causal analytical method for group decision-making under fuzzy environment. Expert Syst. Appl. **34**(1), 205–213 (2008)
13. Organ, A.: Bulanık Dematel Yöntemiyle Makine Seçimini Etkileyen Kriterlerin Değerlendirilmesi. Çukurova Üniversitesi Sosyal Bilimler Enstitüsü Dergisi, vol. 22, no. 1, pp. 157–172, December 2013
14. Aksakal, E., Dağdeviren, M.: ANP Ve DEMATEL Yöntemleri İle Personel Seçimi Problemine Bütünleşik Bir Yaklaşım. Gazi Üniv. Müh. Mim. Fak. Der. vol. 25, no. 4, pp. 905–913 (2010)
15. Yücenur, G.N.: Turizm Sektöründe Strateji Seçimi için Bulanık Veriler Yardımıyla Hiyerarşik Ağ Modeli ve SWOT Analizi: Türkiye Örneği. Sakarya Üniversitesi Fen Bilimleri Enstitüsü Dergisi, vol. 21, no. 5, September 2017
16. Dargi, A., Anjomshoae, A., Galankashi, M.R., Memari, A., Tap, M.B.M.: Supplier selection: a fuzzy-ANP approach. Proc. Comput. Sci. **31**, 691–700 (2014)

17. Chang, D.-Y.: Applications of the extent analysis method on fuzzy AHP. Eur. J. Oper. Res. **95**(3), 649–655 (1996)
18. Le Téno, J.F., Mareschal, B.: An interval version of PROMETHEE for the comparison of building products' design with ill-defined data on environmental quality. Eur. J. Oper. Res. **109**(2), 522–529 (1998)
19. Yılmaz, B., Dağdeviren, M.: Ekipman Seçimi Probleminde PROMETHEE ve Bulanık PROMETHEE Yöntemlerinin Karşılaştırmalı Analizi. Gazi Üniversitesi Mühendislik-Mimarlık Fakültesi Dergisi, vol. 25, no. 4, pp. 811–826 (2010)

Reproduction of Experiments in Recommender Systems Evaluation Based on Explanations

Nikolaos Polatidis[1]([⊠]) and Elias Pimenidis[2]

[1] School of Computing, Engineering, and Mathematics,
University of Brighton, BN2 4GJ Brighton, UK
N.Polatidis@Brighton.ac.uk
[2] Department of Computer Science and Creative Technologies,
University of the West of England, BS16 1QY Bristol, UK
Elias.Pimenidis@uwe.ac.uk

Abstract. The offline evaluation of recommender systems is typically based on accuracy metrics such as the Mean Absolute Error (MAE) and the Root Mean Squared Error (RMSE), while on the other hand Precision and Recall is used to measure the quality of the top-N recommendations. However, it is difficult to reproduce the results since there are different libraries that can be used for running experiments and also within the same library there are many different settings that if not taken into consideration when replicating the result might vary. In this paper, we show that it is challenging to reproduce results using a different library but with the use of the same library an explanation based approach can be used to assist in the reproducibility of experiments. Our proposed approach has been experimentally evaluated using a real dataset and the results show that it is both practical and effective.

Keywords: Recommender systems · Evaluation · Explanations ·
Reproducibility

1   Introduction

Recommender systems are widely known for their use in e-Commerce for recommending products to users, thus reducing the overall searching time of the user and increase sales. Furthermore, it is a technology used in various other less known domains such as music recommendation or people to people recommendation in social media [1]. However, the increasingly use and popularity of recommender systems research both in academia and in industry has lead us to the development of new algorithms and their experimental evaluation. While this is important to do, it should be noted that the problem of reproducing the results exists and it is considered important [2]. For the offline evaluation of recommender systems various metrics can be used such as MAE and RMSE for predicting the accuracy error and information retrieval metrics such as Precision and Recall can be used for measuring the quality of the top-N recommendations [3]. While, there are more metrics it is outside of the scope of this paper to discuss them however further details can be found in [3]. In the literature there are different

© Springer Nature Switzerland AG 2018
E. Pimenidis and C. Jayne (Eds.): EANN 2018, CCIS 893, pp. 194–200, 2018.
https://doi.org/10.1007/978-3-319-98204-5_16

libraries that can be used for developing and testing a recommendation algorithm and include Recommender101, Apache Mahout, LensKit and MyMediaLite among others [1, 4]. In the work by [1] it has been shown that reproducing the experimental results of an algorithm is very difficult when using a different library because of different settings and parameters that exist between them. However, it is shown that if a set of carefully selected guidelines is followed with the use of the same library then the results can be replicated with a very small and non-noticeable different in the output value.

To assist in solving the problem of reproducibility of experiments with the use of the same recommendation library we have:

1. Developed an explanation based approach that can be used towards this direction.
2. Experimentally evaluated the above approach using a publicly available recommendation library and a real dataset.

The rest of the paper is organized as follows: Sect. 2 provides the relevant background, Sect. 3 delivers the proposed approach, Sect. 4 presents the experiments and Sect. 5 contains the conclusions.

2 Background

Evaluating recommender systems in offline environments can be done using prediction accuracy or information retrieval metrics. However, the problem arises when in a research output of a new algorithm the source code is not made publicly available or when the exact settings for replicating the code and the experiments are missing. In the literature there are related works that have done important steps towards the solution of the reproducibility problem. In [1] a very good analysis of the main problems is identified, which include the name and the source code of the recommendation library, the details of the algorithm, the dataset used and the details of how the dataset has been used. Moreover, in the same work a set of guidelines is proposed that can be followed to assist in the reproducibility. Another similar work that identifies a set of best practices for recommender systems can be found in [5], while in [6] the importance of the reproducibility of experiments in recommender systems evaluation was highlighted with the organization of a workshop in 2013. Furthermore, the outcome of this workshop can be found in a relevant report with its future directions being theoretical only [7]. One other relevant approach can be found in [8] and it is about the improvement of statistical power of the 10-fold cross validation scheme in recommender systems. A more relevant but more software oriented approach is Rival [9]. In this approach a toolkit provided different stages in the process such as data splitting, item recommendation and evaluation. It is not however a framework or a library but a toolkit that can be used in Apache Mahout, LensKit and MyMediaLite and it provides a user interface. Other researchers however having knowing about the reproducibility problem have decided to develop and propose their own evaluation metrics. For example, in [10] the authors proposed a general evaluation metric that operate over a set of sessions, while another proposed metric can be found in [11] where the authors propose the modified Reciprocal Hit Rand Metric (mRHR) which is a hit rank metric.

In addition to the related works, the most common recommendation method is Collaborative Filtering (CF) and the most know CF method is Pearson Correlation Coefficient (PCC). PCC is defined in Eq. 1 and Sim (a, b) is the similarity between users a and b, also $r_{a,p}$ is the rating of user a for product p, $r_{b,p}$ is the rating of user b for product p and $\acute{r}a$ and $\acute{r}b$ represent the user's average ratings. P is the set of all products. Moreover, the similarity value ranges from −1 to 1 and higher is better.

$$
\underset{a,b}{PCC} = \frac{\sum p \in P(ra, p - \acute{r}a)(rb, p - \acute{r}b)}{\sqrt{\sum p \in P(ra, p - \acute{r}a)^2}\sqrt{\sum p \in P(rb, p - \acute{r}b)^2}}
\tag{1}
$$

Furthermore, to measure the prediction error, MAE it typically be used and is defined in Eq. 2 where pi is the predicted rating and ri is the actual rating in the summation. This method is used for the computation of the deviation between the predicted ratings and the actual ratings. It should also be noted that lower values are better.

$$
MAE = \frac{1}{n}\sum_{i=1}^{n}|pi - ri|
\tag{2}
$$

However, there are numerous settings found in a recommendation library that can affect the result, such as the number of the nearest neighbors, if a cross-fold evaluation took place or the dataset what split into a training and testing part and the minimum ratings per item or if a threshold of minimum ratings that a user has submitted for an item will be applied. In Table 1 we can see the results of PCC using different neighborhood size, the MovieLens 1 million dataset [12], 80% training and 20% testing evaluation and the Recommender101 library. Furthermore, in Table 2 it is shown that if the minimum number of ratings per user is different the output can vary significantly on a 5-fold cross validation.

Table 1. MAE results for Recommender101

| | Number of k nearest neighbours | | | | | |
	60	80	100	200	300	400
PCC	0.870	0.862	0.841	0.811	0.785	0.761

Table 2. 5-fold cross-fold with different settings MAE results based on Recommender101

Settings	Min number of ratings per user (30)	Min number of ratings per user (Not known and not specified – Default value used by the library)
PCC	0.872	0.890

3 Proposed Approach

In previous research it has been shown that it is very difficult to reproduce results using different evaluation libraries due to differences that exist between the implementations of algorithms and metrics [1]. However, with the use of the same library the possibility of reproducing correctly an algorithm and an experimental evaluation is high if the same settings and parameters are used.

Thus, for our proposed approach we use the Recommender101 library in combination with a set of explanations that accompany the output log file of the result. The library is comprised from a set of components for offline evaluation as shown in Fig. 1. The settings used such as the algorithm used, the number nearest neighbors, type of validation (cross fold or test/train) and the algorithm evaluated are passed in an external configuration file. Moreover, it supports well known metrics such as MAE, RMSE, Precision,

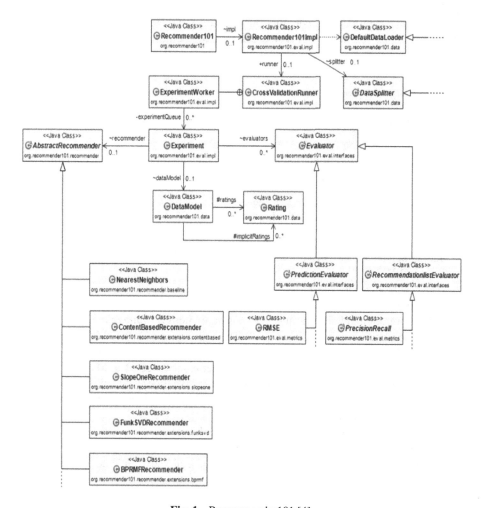

Fig. 1. Recommender101 [4]

Recall, NDCG among others and when the experiment is finished the result is printed on the screen and saved in a log file. We extend the Recommender101 library to print on the screen and also save in the log file a set of explanations in simple language that can be used to guide a future researcher to reproduce an experiment.

In addition to the settings used it should be noted that it is difficult to exactly replicate an experiment since in most cases a dataset is randomly divided to training and testing parts. However and despite of this minor issue if the other settings and parameters are properly applied then the result will be close to identical.

3.1 Explanations

We define explanations as a set of details that accompany the output result, thus making it clear to the research what needs to be included in a research output. In Recommender101 a number of settings and parameters are available in the configuration file (recommender101.properties under the conf directory). If these parameters are not properly mentioned by researchers in their work then the output result could vary significantly [1].

3.2 The Proposed Approach

In the proposed approach we:

1. Retrieve information from the configuration file
2. Write the information in the log file along with evaluation result and explain what this is

The settings retrieved from the configuration file are the following and are presented in the same way that are saved in the log file:

1. The configuration parameters and settings can be set at the configuration file recommender101.properties that be found under the conf directory of Recommender101
2. The filename of the dataset is (name of the file goes here)
3. The minimum number of ratings per user to be considered is (number)
4. The minimum number of ratings per considered item is (number)
5. This experiment has used all users OR This experiments has used (number) users
6. The minimum rating value applied is (number e.g. between 1 to 5)
7. The maximum rating value applied is (number e.g. between 1 to 5)
8. This experiment is based on a (number e.g. 5 or 10) cross fold validation OR this experiment is based on a training/test approach using (number %) for training and (number %) for testing
9. The number of nearest neighbors used is (number)
10. The algorithm used is (name)
11. The metrics used for this experiments are (This is already implemented in recommender101)
12. The results are (This is already implemented in recommender101)

4   Experimental Evaluation

The experimental evaluation has been based on the MovieLens 1 million dataset [12], which consists of 6040 users, 4000 movies and 1 million ratings in a 1–5 scale. Furthermore the Recommender101 library has been used [4]. Furthermore, we have used PCC as the algorithm and MAE as the evaluation metric to perform an experiment with 80% of the dataset used for training and 20% for testing and reproduce the result. Furthermore for each user to be considered a threshold of 20 ratings was applied.

For the experiments we asked two different researchers to perform an experiment each. Both were instructed to download and install Recommender101 in Eclipse. The first one was instructed to perform an experiment and the second one was instructed to use the log file of the first and reproduce the experiment. The results of the first experiment are presented in Table 3 and the results of the second in Table 4. The log file included the MAE result using 100, 200 and 300 k nearest neighbors and all the settings explained in Sect. 3.2.

Table 3. First MAE experiment

	Number of k nearest neighbors		
	100	200	300
PCC	0.841	0.811	0.785

Table 4. Second MAE experiment

	Number of k nearest neighbors		
	100	200	300
PCC	0.842	0.810	0.784

5   Conclusions and Future Work

In this paper we highlighted the problem of reproducibility in recommender systems evaluation. Although, it is shown in previous research that it is difficult to reproduce results using different offline evaluation libraries, the reproducibility of results becomes achievable if the correct settings and parameters are used within the same library. Thus, we have proposed an approach that is based on explanations that can be used to assist researchers in reproducing the results of an experimental evaluation. The initial evaluation results are promising and can assist towards this direction and our approach can be straightforwardly implemented by researchers in other libraries. Furthermore, in our future work we aim to provide a visualized approach of the explanations.

References

1. Polatidis, N., Kapetanakis, S., Pimenidis, E., Kosmidis, K.: Reproducibility of experiments in recommender systems evaluation. In: Iliadis, L., Maglogiannis, I., Plagianakos, V. (eds.) AIAI 2018. IAICT, vol. 519, pp. 401–409. Springer, Cham (2018). https://doi.org/10.1007/978-3-319-92007-8_34
2. Said, A., Bellogín, A.: Comparative recommender system evaluation. In: Proceedings of the 8th ACM Conference on Recommendation Systems - RecSys 2014, pp. 129–136 (2014)
3. Herlocker, J.L., Konstan, J.A., Terveen, L.G., Riedl, J.T.: Evaluating collaborative filtering recommender systems. ACM Trans. Inf. Syst. **22**, 5–53 (2004)
4. Jannach, D., Lerche, L., Gedikli, F., Bonnin, G.: What recommenders recommend – an analysis of accuracy, popularity, and sales diversity effects. In: Carberry, S., Weibelzahl, S., Micarelli, A., Semeraro, G. (eds.) UMAP 2013. LNCS, vol. 7899, pp. 25–37. Springer, Heidelberg (2013). https://doi.org/10.1007/978-3-642-38844-6_3
5. Konstan, J.A., Adomavicius, G.: Toward identification and adoption of best practices in algorithmic recommender systems research. In: Proceedings of the International Workshop on Reproducibility and Replication in Recommender Systems Evaluation, pp. 23–28. ACM, New York (2013)
6. Bellogin, A., Castells, P., Said, A., Tikk, D.: Workshop on reproducibility and replication in recommender systems evaluation. In: Proceedings of the 7th ACM conference on Recommender systems - RecSys 2013, pp. 485–486 (2013)
7. Bellogin, A., Castells, P., Said, A., Tikk, D.: Report on the workshop on reproducibility and replication in recommender systems evaluation (RepSys). SIGIR Forum. **48**, 29–35 (2014)
8. Košir, A., Odić, A., Tkalčič, M.: How to improve the statistical power of the 10-fold cross validation scheme in recommender systems. In: RecSys RepSys 2013: Proceedings of the International Workshop on Reproducibility and Replication in Recommender Systems Evaluation, pp. 3–6 (2013)
9. Said, A., Bellogín, A.: RiVal – a toolkit to foster reproducibility in recommender system evaluation. In: RecSys 2014 Proceedings of the 8th ACM Conference on Recommendation Systems, pp. 371–372 (2014)
10. Hernández del Olmo, F., Gaudioso, E.: Evaluation of recommender systems: a new approach. Expert Syst. Appl. **35**, 790–804 (2008)
11. Peker, S., Kocyigit, A.: mRHR: a modified reciprocal hit rank metric for ranking evaluation of multiple preferences in Top-N recommender systems. In: Dichev, C., Agre, G. (eds.) AIMSA 2016. LNCS (LNAI), vol. 9883, pp. 320–329. Springer, Cham (2016). https://doi.org/10.1007/978-3-319-44748-3_31
12. Harper, F.M., Konstan, J.A.: The MovieLens datasets. ACM Trans. Interact. Intell. Syst. **5**, 1–19 (2015)

Recurrent Neural Networks and Spiking Neural Networks

Recurrent Auto-Encoder Model for Large-Scale Industrial Sensor Signal Analysis

Timothy Wong[(✉)] [iD] and Zhiyuan Luo [iD]

Royal Holloway, University of London, Egham TW20 0EX, UK
timothy.wong@hotmail.co.uk

Abstract. Recurrent auto-encoder model summarises sequential data through an encoder structure into a fixed-length vector and then reconstructs the original sequence through the decoder structure. The summarised vector can be used to represent time series features. In this paper, we propose relaxing the dimensionality of the decoder output so that it performs partial reconstruction. The fixed-length vector therefore represents features in the selected dimensions only. In addition, we propose using rolling fixed window approach to generate training samples from unbounded time series data. The change of time series features over time can be summarised as a smooth trajectory path. The fixed-length vectors are further analysed using additional visualisation and unsupervised clustering techniques. The proposed method can be applied in large-scale industrial processes for sensors signal analysis purpose, where clusters of the vector representations can reflect the operating states of the industrial system.

Keywords: Recurrent auto-encoder · Multidimensional time series
Industrial sensors · Signal analysis

1 Background

Modern industrial processes are often monitored by a large array of sensors. Machine learning techniques can be used to analyse unbounded streams of sensor signal in an on-line scenario. This paper illustrates the idea using proprietary data collected from a two-stage centrifugal compression train driven by an aeroderivative industrial engine (Rolls-Royce RB211) on a single shaft. This large-scale compression module belongs to a major natural gas terminal[1]. The purpose of this modular process is to regulate the pressure of natural gas at an

---

Supported by Centrica plc. Registered office: Millstream, Maidenhead Road, Windsor SL4 5GD, United Kingdom.

[1] A simplified process diagram of the compression train can be found in Fig. 6 at the appendix.

© Springer Nature Switzerland AG 2018
E. Pimenidis and C. Jayne (Eds.): EANN 2018, CCIS 893, pp. 203–216, 2018.
https://doi.org/10.1007/978-3-319-98204-5_17

elevated, pre-set level. At the compression system, sensors are installed to monitor the production process. Real-valued measurements such as temperature, pressure, rotary speed, vibration... etc., are recorded at different locations[2].

Streams of sensor signals can be treated as a multidimensional entity changing through time. Each stream of sensor measurement is basically a set of real values received in a time-ordered fashion. When this concept is extended to a process with P sensors, the dataset can therefore be expressed as a time-ordered multidimensional vector $\{\mathbb{R}_t^P : t \in [1, T]\}$.

The dataset used in this study is unbounded (i.e. continuous streaming) and unlabelled, where the events of interest (e.g. overheating, mechanical failure, blocked oil filters... etc) are not present. The key goal of this study is to identify sensor patterns and anomalies to assist equipment maintenance. This can be achieved by finding the representation of multiple sensor data. We propose using recurrent auto-encoder model to extract vector representation for multidimensional time series data. Vectors can be analysed further using visualisation and clustering techniques in order to identify patterns.

1.1 Related Works

A comprehensive review [1] analysed traditional clustering algorithms for unidimensional time series data. It has concluded that Dynamic Time Warping (DTW) can be an effective benchmark for unidimensional time series data representation. There has been attempts to generalise DTW to multidimensional level [5,6,8,11,13,15,16,20,21]. Most of these studies focused on analysing time series data with relatively low dimensionality, such as those collected from Internet of Things (IoT) devices, wearable sensors and gesture recognition. This paper contributes further by featuring a time series dataset with much higher dimensionality which is representative for any large-scale industrial applications.

Among neural network researches, [18] proposed a recurrent auto-encoder model based on LSTM neurons which aims at learning video data representation. It achieves this by reconstructing sequence of video frames. Their model was able to derive meaningful representations for video clips and the reconstructed outputs demonstrate sufficient similarity based on qualitative examination. Another recent paper [4] also used LSTM-based recurrent auto-encoder model for video data representation. Sequence of frames feed into the model so that it learns the intrinsic representation of the underlying video source. Areas with high reconstruction error indicate divergence from the known source and hence can be used as a video forgery detection mechanism.

Similarly, audio clips can treated as sequential data. A study [3] converted variable-length audio data into fixed-length vector representation using recurrent auto-encoder model. It found that audio segments that sound alike usually have vector representations in same neighbourhood.

There are other works related to time series data. For instance, a recent paper [14] proposed a recurrent auto-encoder model which aims at providing fixed-

---

[2] A list of sensors is available in the appendix.

length representation for bounded univariate time series data. The model was trained on a plurality of labelled datasets with the aim of becoming a generic time series feature extractor. Dimensionality reduction of the vector representation via t-SNE shows that the ground truth labels can be observed in the extracted representations. Another study [9] proposed a time series compression algorithm using a pair of RNN encoder-decoder structure and an additional auto-encoder to achieve higher compression ratio. Meanwhile, another research [12] used an auto-encoder model with database metrics (e.g. CPU usage, number of active sessions... etc) to identify anomalous usage periods by setting threshold on the reconstruction error.

2   Methods

A pair of RNN encoder-decoder structure can provide end-to-end mapping between an ordered multidimensional input sequence and its matching output sequence [2,19]. Recurrent auto-encoder can be depicted as a special case of the aforementioned model, where input and output sequences are aligned with each other. It can be extended to the area of signal analysis in order to leverage recurrent neurons power to understand complex and time-dependent relationship.

2.1   Encoder-Decoder Structure

At high level, the RNN encoder reads an input sequence and summarises all information into a fixed-length vector. The decoder then reads the vector and reconstructs the original sequence. Figure 1 below illustrates the model.

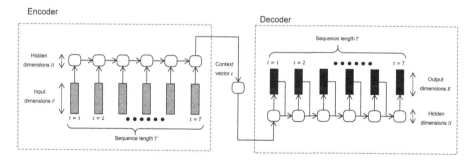

Fig. 1. Recurrent auto-encoder model. Both the encoder and decoder are made up of multilayered RNN. Arrows indicate the direction of information flow.

Encoding. The role of the recurrent encoder is to project the multidimensional input sequence into a fixed-length hidden context vector c. It reads the input vectors $\{\mathbb{R}_t^P : t \in [1, T]\}$ sequentially from $t = 1, 2, 3, ..., T$. The hidden state of the RNN has H dimensions which updates at every time step based on the current input and hidden state inherited from previous steps.

Recurrent neurons arranged in multiple layers are capable of learning complex temporal behaviours. In this proposed model, LSTM neurons with hyperbolic tangent activation are used at all recurrent layers [7]. An alternative choice of using gated recurrent unit (GRU) neurons [2] can also be used but was not experimented within the scope of this study. Once the encoder reads all the input information, the sequence is summarised in a fixed-length vector c which has H hidden dimensions.

For regularisation purpose, dropout can be applied to avoid overfitting. It refers to randomly removing a fraction of neurons during training, which aims at making the network more generalisable [17]. In an RNN setting, [22] suggested that dropout should only be applied non-recurrent connections. This helps the recurrent neurons to retain memory through time while still allowing the non-recurrent connections to benefit from regularisation.

Decoding. The decoder is a recurrent network which uses the representation c to reconstruct the original sequence. To exemplify this, the decoder starts by reading the context vector c at $t = 1$. It then decodes the information through the RNN structure and outputs a sequence of vectors $\{\mathbb{R}_t^K : t \in [1, T]\}$ where K denotes the dimensionality of the output sequence.

Recalling one of the fundamental characteristics of an auto-encoder is the ability to reconstruct the input data back into itself via a pair of encoder-decoder structure. This criterion can be slightly relaxed such that $K \leqslant P$, which means the output sequence is only a partial reconstruction of the input sequence.

Recurrent auto-encoder with partial reconstruction:

$$\begin{cases} f_{encoder} : \{\mathbb{R}_t^P : t \in [1, T]\} \to c \\ f_{decoder} : c \to \{\mathbb{R}_t^K : t \in [1, T]\} \end{cases} \quad K \leqslant P \tag{1}$$

In the large-scale industrial system use case, all streams of sensor measurements are included in the input dimensions while only a subset of sensors is included in the output dimensions. This means that the entire system is visible to the encoder, but the decoder only needs to perform partial reconstruction of it. End-to-end training of the relaxed auto-encoder implies that the context vector would summarise the input sequence while still being conditioned on the output sequence. Given that activation of the context vector is conditional on the decoder output, this approach allows the encoder to capture lead variables across the entire process as long as they are relevant to the selected output dimensions.

It is important to recognise that reconstructing part of the data is an easier task to perform than fully-reconstructing the entire original sequence. However, partial reconstruction has practical significance for industrial applications. In real-life scenarios, multiple context vectors can be generated from different recurrent auto-encoder models using identical sensors in the encoder input but different subset of sensors in the decoder output. The selected subsets of sensors can reflect the underlying operating states of different parts of the industrial system. As a result, context vectors produced from the same temporal segment can be used as different diagnostic measurements in industrial context. We will illustrate this in the results section by two examples.

2.2 Sampling

For a training dataset of T' time steps, samples can be generated where $T < T'$. We can begin at $t = 1$ and draw a sample of length T. This process continues recursively by shifting one time step until it reaches the end of the training dataset. For a subset sequence with length T, this method allows $T' - T$ samples to be generated. Besides, it can also generate samples from an unbounded time series in an on-line scenario, which are essential for time-critical applications such as sensor data analysis.

Algorithm 1. Drawing samples consecutively from the original dataset

---

Input: Dataset length T'
Input: Sample length T
1 $i \leftarrow 0$;
2 **while** $i \leqslant i + T$ **do**
3 | Generate sample sequence $(i, i + T]$ from the dataset;
4 | $i \leftarrow i + 1$;
5 **end**

Given that sample sequences are recursively generated by shifting the window by one time step, successively-generated sequences are highly correlated with each other. As we have discussed previously, the RNN encoder structure compresses sequential data into a fixed-length vector representation. This means that when consecutive sequences are fed through the encoder structure, the resulting activation at c would also be highly correlated. As a result, consecutive context vectors can join up to form a smooth trajectory in space.

Context vectors in the same neighbourhood have similar activation therefore the industrial system must have similar underlying operating states. Contrarily, context vectors located in distant neighbourhoods would have different underlying operating states. These context vectors can be visualised in lower dimensions via dimensionality reduction techniques such as principal component analysis (PCA).

Furthermore, additional unsupervised clustering algorithms can be applied to the context vectors. Each context vector can be assigned to a cluster C_j where J is the total number of clusters. Once all the context vectors are labelled with their corresponding clusters, supervised classification algorithms can be used to learn the relationship between them using the training set. For instance, support vector machine (SVM) classifier with J classes can be used. The trained classifier can then be applied to the context vectors in the held-out validation set for cluster assignment. It can also be applied to context vectors generated from unbounded time series in an on-line setting. Change in cluster assignment among successive context vectors indicates a change in the underlying operating state.

3   Results

Training samples were drawn from the dataset using windowing approach with fixed sequence length. In our example, the large-scale industrial system has 158 sensors which means the recurrent auto-encoder's input dimension has $P = 158$. Observations are taken at 5 min granularity and the total duration of each sequence was set at 3 h. This means that the model's sequence has fixed length $T = 36$, while samples were drawn from the dataset with total length $T' = 2724$. The dataset was scaled into z-scores, thus ensuring zero-centred data which facilitates gradient-based training.

The recurrent auto-encoder model has three layers in the RNN encoder structure and another three layers in the corresponding RNN decoder. There are 400 neurons in each layer. The auto-encoder model structure can be summarised as: RNN encoder (400 neurons/3 layers LSTM/hyperbolic tangent) - Context layer (400 neurons/Dense/linear activation) - RNN decoder (400 neurons/ 3 layers LSTM/hyperbolic tangent). Adam optimiser [10] with 0.4 dropout rate was used for model training.

3.1   Output Dimensionity

As we discussed earlier, the RNN decoder's output dimension can be relaxed for partial reconstruction. The output dimensionality was set at $K = 6$ which is comprised of a selected set of sensors relating to key pressure measurements (e.g. suction and discharge pressures of the compressor device).

We have experimented three scenarios where the first two have complete dimensionality $P = 158; K = 158$ and $P = 6; K = 6$ while the remaining scenario has relaxed dimensionality $P = 158; K = 6$. The training and validation MSEs of these models are visualised in Fig. 2 below.

The first model with complete dimensionality ($P = 158; K = 158$) has visibility of all dimensions in both the encoder and decoder structures. Yet, both the training and validation MSEs are high as the model struggles to compress-decompress the high dimensional time series data.

For the complete dimensionality model with $P = 6; K = 6$, the model has limited visibility to the system as only the selected dimensions were included. Despite the context layer summarises information specific to the selected dimensionality in this case, lead variables in the original dimensions have been excluded. This prevents the model from learning any dependent behaviours among all available information.

On the other hand, the model with partial reconstruction ($P = 158; K = 6$) demonstrate substantially lower training and validation MSEs. Since all information is available to the model via the RNN encoder, it captures the relevant information such as lead variables across the entire system.

Randomly selected samples in the held-out validation set were fed to this model and the predictions can be qualitatively examined in details. In Fig. 3 below, all the selected specimens demonstrate high similarity between the original label and the reconstructed output. The recurrent auto-encoder model captures the shift in mean level as well as temporal variations across all output dimensions.

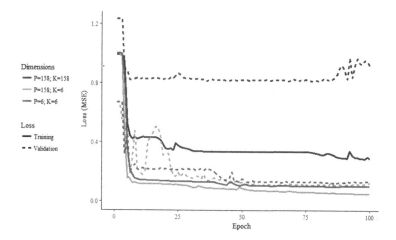

Fig. 2. Effects of relaxing dimensionality of the output sequence on the training and validation MSE losses. They contain same number of layers in the RNN encoder and decoder respectively. All hidden layers contain same number of LSTM neurons with hyperbolic tangent activation.

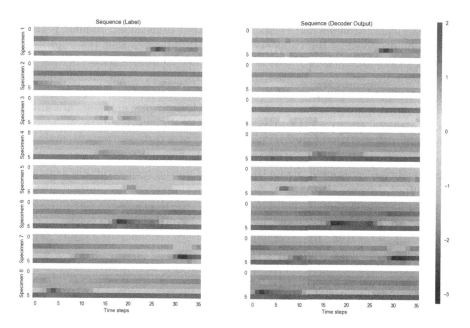

Fig. 3. A heatmap showing eight randomly selected output sequences in the held-out validation set. Colour represents magnitude of sensor measurements in normalised scale.

3.2   Context Vector

Once the recurrent auto-encoder model is successfully trained, samples can be fed
to the model and the corresponding context vectors can be extracted for detailed
inspection. In the model we selected, the context vector c is a multi-dimensional
real vector $\mathbb{R}^{400}$. Since the model has input dimensions $P = 158$ and sequence
length $T = 36$, the model has achieved compression ratio $\frac{158 \times 36}{400} = 14.22$. Dimensionality reduction of the context vectors through principal component analysis
(PCA) shows that context vectors can be efficiently embedded in lower dimensions (e.g. two-dimensional space).

At low-dimensional space, we used supervised classification algorithm to learn
the relationship between vectors representations and cluster assignment. The
trained classification model can then be applied to the validation set to assign
clusters for unseen data. In our experiment, a SVM classifier with radial basis
function (RBF) kernel ($\gamma = 4$) was used. The results are shown in Fig. 4 below.

In two-dimensional space, the context vectors separate into two clearly identifiable neighbourhoods. These two distinct neighbourhoods correspond to the
shift in mean values across all output dimensions. When K-means clustering
algorithm is applied, it captures these two neighbourhoods as two clusters in the
scenario depicted in Fig. 4a.

When the number of clusters increases, they begin to capture more subtleties.
In the six clusters scenario illustrated in Fig. 4b, successive context vectors oscil-
late back and forth between neighbouring clusters. The trajectory corresponds
to the interlacing troughs and crests in the output dimensions. Similar pattern
can also be observed in the validation set, which indicates that the knowledge
learned by the auto-encoder model is generalisable to unseen data.

Furthermore, we have repeated the same experiment again with a different
configuration ($K = 158; P = 2$) to reassure that the proposed approach can
provide robust representations of the data. The sensor measurements are drawn
from an identical time period and only the output dimensionality K is changed
(The newly selected set of sensors is comprised of a different measurements of
discharge gas pressure at the compressor unit). Through changing the output
dimensionality K, we can illustrate the effects of partial reconstruction using different output dimensions. As seen in Fig. 5, the context vectors form a smooth
trajectory in the low-dimensional space. Similar sequences yield context vectors
which are located in a shared neighbourhood. Nevertheless, the clusters found
by K-means method in this secondary example also manage to identify neighbourhoods with similar sensor patterns.

4   Discussion and Conclusion

Successive context vectors generated by windowing approach are always highly
correlated, thus form a smooth trajectory in high-dimensional space. Additional
dimensionality reduction techniques can be applied to visualise the change of
time series features. One of the key contributions of this study is that similar

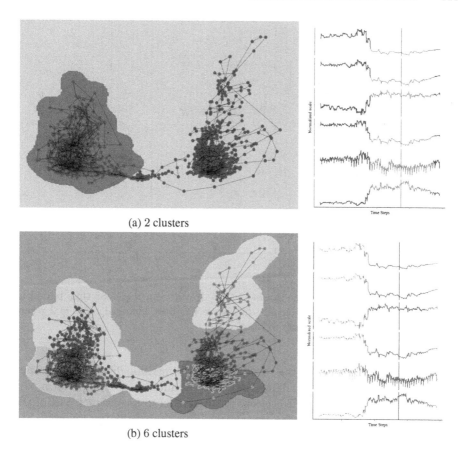

(a) 2 clusters

(b) 6 clusters

Fig. 4. The first example. On the left, the context vectors were projected into two-dimensional space using PCA. The black solid line on the left joins all consecutive context vectors together as a trajectory. Different number of clusters were identified using simple K-means algorithm. Cluster assignment and the SVM decision boundaries are coloured in the charts. On the right, output dimensions are visualised on a shared time axis. The black solid line demarcates the training set (70%) and validation sets (30%). The line segments are colour-coded to match the corresponding clusters.

context vectors can be grouped into clusters using unsupervised clustering algorithms such as K-means algorithm. Clusters can be optionally labelled manually to identify operating state (e.g. healthy vs. faulty). Alarm can be triggered when the context vector travels beyond the boundary of a predefined neighbourhood. Clusters of the vector representation can be used by operators and engineers to aid diagnostics and maintenance.

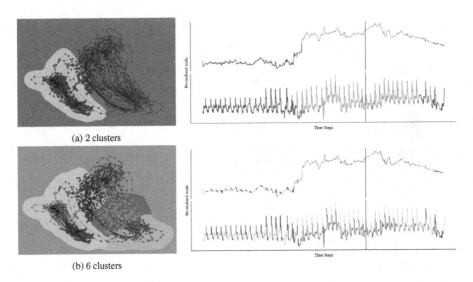

(a) 2 clusters

(b) 6 clusters

Fig. 5. The second example. The sensor data is drawn from the same time period as the previous example, only the output dimension has been changed to $K = 2$ where another set of gas pressure sensors were selected.

Another contribution of this study is that dimensionality of the output sequence can be relaxed. This allows the recurrent auto-encoder to perform partial reconstruction. Although it is easier for the model to reconstruct part of the original sequence, such simple improvement allows users to define different sets of sensors of particular interest. By changing sensors in the decoder output, context vectors can be used to reflect underlying operating states of various aspects of the large-scale industrial process. This ultimately enables users to diagnose the industrial system by generating more useful insight.

This proposed method essentially performs multidimensional time series clustering. We have demonstrated that it can natively scale up to very high dimensionality as it is based on recurrent auto-encoder model. We have applied the method to an industrial sensor dataset with $P = 158$ and empirically show that it can represent multidimensional time series data effectively. In general, this method can be further generalised to any multi-sensor multi-state processes for operating state recognition.

This study established that recurrent auto-encoder model can be used to analyse unlabelled and unbounded time series data. It further demontrated that operating state (i.e. labels) can be inferred from unlabelled time series data. This opens up further possibilities for analysing complex industrial sensors data given that it is predominately overwhelmed with unbounded and unlabelled time series data.

Nevertheless, the proposed approach has not included any categorical sensor measurements (e.g. open/closed, tripped/healthy, start/stop... etc). Future research can focus on incorporating categorical measurements alongside real-valued measurements.

Disclosure

The technical method described in this paper is the subject of British patent application GB1717651.2.

Appendix A

The rotary components are driven by industrial RB-211 jet turbine on a single shaft through a gearbox. Incoming natural gas passes through the low pressure (LP) stage first which brings it to an intermediate pressure level, it then passes through the high pressure (HP) stage and reaches the pre-set desired pressure level. The purpose of the suction scrubber is to remove any remaining condensate from the gas prior to feeding through the centrifugal compressors. Once the hot compressed gas is discharged from the compressor, its temperature is lowered via the intercoolers (Fig. 7).

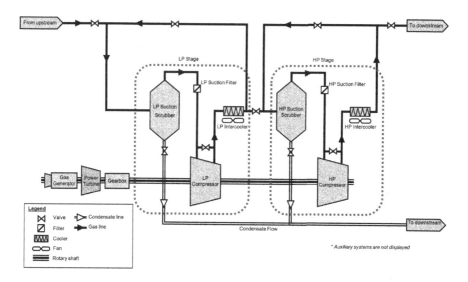

Fig. 6. A simplified process diagram of the two-stage centrifugal compression train which is located at a natural gas terminal.

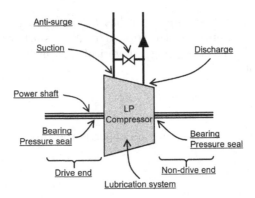

Fig. 7. Locations of key components around the centrifugal compressor.

Appendix B

The sensor measurements used in the analysis are listed below:

1. GASCOMPCARBONDIOXIDEMEAS
2. GASCOMPMETHANEMEAS
3. GASCOMPNITROGENMEAS
4. GASPROPMOLWTMEAS
5. PRESSAMBIENT
6. GB_SPEEDINPUT
7. GB_SPEEDOUTPUT
8. GB_TEMPINPUTBRGDRIVEEND
9. GB_TEMPINPUTBRGNONDRIVEEND
10. GB_TEMPINPUTBRGTHRUSTINBOARD
11. GB_TEMPINPUTBRGTHRUSTOUTBRD
12. GB_TEMPLUBOIL
13. GB_TEMPLUBOILTANK
14. GB_TEMPOUTPUTBRGDRIVEEND
15. GB_TEMPOUTPUTBRGNONDRIVEEND
16. GB_VIBBRGCASINGVEL
17. GB_VIBINPUTAXIALDISP
18. GB_VIBINPUTDRIVEEND
19. GB_VIBINPUTNONDRIVEEND
20. GB_VIBOUTPUTDRIVEEND
21. GB_VIBOUTPUTNONDRIVEEND
22. GG_FLOWFUEL
23. GG_FLOWWATERINJECTION
24. GG_FLOWWATERINJSETPOINT
25. GG_POWERSHAFT
26. GG_PRESSAIRINLET
27. GG_PRESSCOMPDEL
28. GG_PRESSCOMPDELHP
29. GG_PRESSCOMPDELIP
30. GG_PRESSDIFBRGLUBOIL
31. GG_PRESSDIFINLETFILTER
32. GG_PRESSDIFINLETFLARE
33. GG_PRESSDIFVALVEWATERINJCTRL
34. GG_PRESSDISCHWATERINJPUMP1
35. GG_PRESSDISCHWATERINJPUMP2
36. GG_PRESSEXH
37. GG_PRESSFUELGAS
38. GG_PRESSHYDOILDEL
39. GG_PRESSLUBEOILHEADER
40. GG_PRESSLUBOIL
41. GG_PRESSMANIFOLDWATERINJ
42. GG_PRESSSUCTWATERINJPUMP
43. GG_SPEEDHP
44. GG_SPEEDIP
45. GG_TEMPAIRINLET
46. GG_TEMPCOMPDEL
47. GG_TEMPCOMPDELHP
48. GG_TEMPCOMPDELIP
49. GG_TEMPEXH
50. GG_TEMPEXHTC1
51. GG_TEMPEXHTC2
52. GG_TEMPEXHTC3
53. GG_TEMPEXHTC4
54. GG_TEMPEXHTC5
55. GG_TEMPEXHTC6
56. GG_TEMPEXHTC7
57. GG_TEMPEXHTC8
58. GG_TEMPFUELGAS
59. GG_TEMPFUELGASG1
60. GG_TEMPFUELGASLINE
61. GG_TEMPHSOILCOOLANTRETURN
62. GG_TEMPHSOILMAINRETURN
63. GG_TEMPLUBOIL
64. GG_TEMPLUBOILTANK
65. GG_TEMPPURGEMUFF
66. GG_TEMPWATERINJSUPPLY
67. GG_VALVEWATERINJECTCONTROL
68. GG_VANEINLETGUIDEANGLE
69. GG_VANEINLETGUIDEANGLE1
70. GG_VANEINLETGUIDEANGLE2
71. GG_VIBCENTREBRG
72. GG_VIBFRONTBRG
73. GG_VIBREARBRG
74. HP_HEADANTISURGE
75. HP_POWERSHAFT
76. HP_PRESSCLEANGAS
77. HP_PRESSDIFANTISURGE
78. HP_PRESSDIFSUCTSTRAINER
79. HP_PRESSDISCH
80. HP_PRESSSEALDRYGAS
81. HP_PRESSSEALLEAKPRIMARYDE1
82. HP_PRESSSEALLEAKPRIMARYDE2
83. HP_PRESSSEALLEAKPRIMARYNDE1
84. HP_PRESSSEALLEAKPRIMARYNDE2
85. HP_PRESSSUCT1
86. HP_PRESSSUCT2
87. HP_SPEED
88. HP_TEMPBRGDRIVEEND
89. HP_TEMPBRGNONDRIVEEND
90. HP_TEMPBRGTHRUSTINBOARD
91. HP_TEMPBRGTHRUSTOUTBRD
92. HP_TEMPDISCH1
93. HP_TEMPDISCH2
94. HP_TEMPLUBOIL
95. HP_TEMPLUBOILTANK
96. HP_TEMPSUCT1
97. HP_VIBAXIALDISP1
98. HP_VIBAXIALDISP2
99. HP_VIBDRIVEEND
100. HP_VIBDRIVEENDX
101. HP_VIBDRIVEENDY
102. HP_VIBNONDRIVEEND
103. HP_VIBNONDRIVEENDX
104. HP_VIBNONDRIVEENDY
105. HP_VOLDISCH
106. HP_VOLRATIO
107. HP_VOLSUCT
108. LP_HEADANTISURGE
109. LP_POWERSHAFT
110. LP_PRESSCLEANGAS
111. LP_PRESSDIFANTISURGE
112. LP_PRESSDIFSUCTSTRAINER
113. LP_PRESSDISCH
114. LP_PRESSSEALDRYGAS
115. LP_PRESSSEALLEAKPRIMARYDE1
116. LP_PRESSSEALLEAKPRIMARYDE2
117. LP_PRESSSEALLEAKPRIMARYNDE1
118. LP_PRESSSEALLEAKPRIMARYNDE2
119. LP_PRESSSUCT1
120. LP_PRESSSUCT2
121. LP_SPEED
122. LP_TEMPBRGDRIVEEND
123. LP_TEMPBRGNONDRIVEEND
124. LP_TEMPBRGTHRUSTINBOARD
125. LP_TEMPBRGTHRUSTOUTBOARD
126. LP_TEMPDISCH1
127. LP_TEMPDISCH2
128. LP_TEMPLUBOIL
129. LP_TEMPLUBOILTANK
130. LP_TEMPSUCT1
131. LP_VIBAXIALDISP1
132. LP_VIBAXIALDISP2
133. LP_VIBDRIVEEND
134. LP_VIBDRIVEENDX
135. LP_VIBDRIVEENDY
136. LP_VIBNONDRIVEEND
137. LP_VIBNONDRIVEENDX
138. LP_VIBNONDRIVEENDY
139. LP_VOLDISCH
140. LP_VOLRATIO
141. LP_VOLSUCT
142. PT_POWERSHAFT
143. PT_SPEED
144. PT_TEMPBRGDRIVEEND
145. PT_TEMPBRGNONDRIVEEND
146. PT_TEMPBRGTHRUST1
147. PT_TEMPBRGTHRUST3
148. PT_TEMPCOOLINGAIR1
149. PT_TEMPCOOLINGAIR2
150. PT_TEMPEXH
151. PT_TEMPLUBOIL
152. PT_TEMPLUBOILPTSUMP
153. PT_TEMPLUBOILTANK
154. PT_VIBAXIALDISP1
155. PT_VIBAXIALDISP2
156. PT_VIBBRGCASINGVEL
157. PT_VIBDRIVEEND
158. PT_VIBNONDRIVEEND

References

1. Bagnall, A., Lines, J., Bostrom, A., Large, J., Keogh, E.: The great time series classification bake off: a review and experimental evaluation of recent algorithmic advances. Data Min. Knowl. Discov. **31**(3), 606–660 (2017). https://doi.org/10.1007/s10618-016-0483-9
2. Cho, K., van Merrienboer, B., Gülçehre, Ç., Bougares, F., Schwenk, H., Bengio, Y.: Learning phrase representations using RNN encoder-decoder for statistical machine translation. CoRR abs/1406.1078 (2014). http://arxiv.org/abs/1406.1078
3. Chung, Y., Wu, C., Shen, C., Lee, H., Lee, L.: Audio word2vec: Unsupervised learning of audio segment representations using sequence-to-sequence autoencoder. CoRR abs/1603.00982 (2016). http://arxiv.org/abs/1603.00982
4. D'Avino, D., Cozzolino, D., Poggi, G., Verdoliva, L.: Autoencoder with recurrent neural networks for video forgery detection. CoRR abs/1708.08754 (2017). http://arxiv.org/abs/1708.08754
5. Gillian, N.E., Knapp, R.B., O'Modhrain, M.S.: Recognition of multivariate temporal musical gestures using n-dimensional dynamic time warping. In: NIME (2011)
6. Giorgino, T.: Computing and visualizing dynamic time warping alignments in R: the dtw package. J. Stat. Softw. **31**(7), 1–24 (2009). https://doi.org/10.18637/jss.v031.i07
7. Hochreiter, S., Schmidhuber, J.: Long short-term memory. Neural Comput. **9**(8), 1735–1780 (1997). https://doi.org/10.1162/neco.1997.9.8.1735
8. ten Holt, G., Reinders, M., Hendriks, E.: Multi-dimensional dynamic time warping for gesture recognition (2007)
9. Hsu, D.: Time series compression based on adaptive piecewise recurrent autoencoder. CoRR abs/1707.07961 (2017). http://arxiv.org/abs/1707.07961
10. Kingma, D.P., Ba, J.: Adam: A method for stochastic optimization. CoRR abs/1412.6980 (2014). http://arxiv.org/abs/1412.6980
11. Ko, M.H., West, G., Venkatesh, S., Kumar, M.: Online context recognition in multisensor systems using dynamic time warping. In: 2005 International Conference on Intelligent Sensors, Sensor Networks and Information Processing, pp. 283–288, December 2005. https://doi.org/10.1109/ISSNIP.2005.1595593
12. Lee, D.: Anomaly Detection in Multivariate Non-stationary Time Series for Automatic DBMS Diagnosis. ArXiv e-prints, August 2017
13. Liu, J., Wang, Z., Zhong, L., Wickramasuriya, J., Vasudevan, V.: uWave: accelerometer-based personalized gesture recognition and its applications. In: 2009 IEEE International Conference on Pervasive Computing and Communications, pp. 1–9, March 2009. https://doi.org/10.1109/PERCOM.2009.4912759
14. Malhotra, P., TV, V., Vig, L., Agarwal, P., Shroff, G.: TimeNet: pre-trained deep recurrent neural network for time series classification. CoRR abs/1706.08838 (2017). http://arxiv.org/abs/1706.08838
15. Petitjean, F., Inglada, J., Gancarski, P.: Satellite image time series analysis under time warping. IEEE Trans. Geosci. Remote Sens. **50**(8), 3081–3095 (2012). https://doi.org/10.1109/TGRS.2011.2179050
16. Shokoohi-Yekta, M., Hu, B., Jin, H., Wang, J., Keogh, E.: Generalizing DTW to the multi-dimensional case requires an adaptive approach. Data Min. Knowl. Discov. **31**(1), 1–31 (2017). https://doi.org/10.1007/s10618-016-0455-0
17. Srivastava, N., Hinton, G., Krizhevsky, A., Sutskever, I., Salakhutdinov, R.: Dropout: a simple way to prevent neural networks from overfitting. J. Mach. Learn. Res. **15**, 1929–1958 (2014). http://jmlr.org/papers/v15/srivastava14a.html

18. Srivastava, N., Mansimov, E., Salakhutdinov, R.: Unsupervised learning of video representations using lstms. CoRR abs/1502.04681 (2015). http://arxiv.org/abs/1502.04681
19. Sutskever, I., Vinyals, O., Le, Q.V.: Sequence to sequence learning with neural networks. CoRR abs/1409.3215 (2014). http://arxiv.org/abs/1409.3215
20. Vlachos, M., Hadjieleftheriou, M., Gunopulos, D., Keogh, E.: Indexing multidimensional time-series. VLDB J. **15**(1), 1–20 (2006). https://doi.org/10.1007/s00778-004-0144-2
21. Wang, J., Balasubramanian, A., Mojica de la Vega, L., Green, J., Samal, A., Prabhakaran, B.: Word recognition from continuous articulatory movement time-series data using symbolic representations (2013)
22. Zaremba, W., Sutskever, I., Vinyals, O.: Recurrent neural network regularization. CoRR abs/1409.2329 (2014). http://arxiv.org/abs/1409.2329

Deep Neural Networks for Prediction of Exacerbations of Patients with Chronic Obstructive Pulmonary Disease

Vimala Nunavath[1]([✉]), Morten Goodwin[1], Jahn Thomas Fidje[1], and Carl Erik Moe[2]

[1] Centre for Artificial Intelligence Research (CAIR), Department of ICT, University of Agder, Grimstad, Norway
[2] Centre for E-Health Research, Department of IS, University of Agder, Kristiansand, Norway
{vimala.nunavath,morten.goodwin,jahn.t.fidje,carl.e.moe}@uia.no

Abstract. Chronic Obstructive Pulmonary Disease (COPD) patients need help in daily life situations as they are burdened with frequent risks of acute exacerbation and loss of control. An automated monitoring system could lead to timely treatments and avoid unnecessary hospital (re-)admissions and home visits by doctors or nurses. Therefore we present a Deep Artificial Neural Networks for approach prediction of exacerbations, particularly Feed-Forward Neural Networks (FFNN) for classification of COPD patients category and Long Short-Term Memory (LSTM), for early prediction of COPD exacerbations and subsequent triage. The FFNN and LSTM models are trained on data collected from remote monitoring of 94 patients through a real monitoring session and therefore represents realistic home monitoring situations. Most deep learning models require large datasets in order to predict with a high degree of accuracy. Our experiments show that with only 94 patients, the FFNN model is able to reproduce health condition provided by a medical doctor with an accuracy of 92.86% and the LSTM model able to predict COPD patients' health conditions one-day ahead with an accuracy of 84.12%. Based on our results, we believe that our work will help the medical doctors and nurses in identifying patients with acute exacerbation in advance which can lead to better patient care and decision making, and hence reduction of costs.

Keywords: COPD · Deep recurrent neural networks
Classification and prediction · LSTM · ANN
Feed-Forward Neural Networks · Support vector machine

1 Introduction

Millions of people are diagnosed every year with a chest disease world wide, and Chronic Obstructive Pulmonary Disease (COPD) is one of the most often

© Springer Nature Switzerland AG 2018
E. Pimenidis and C. Jayne (Eds.): EANN 2018, CCIS 893, pp. 217–228, 2018.
https://doi.org/10.1007/978-3-319-98204-5_18

diagnosed chest diseases [8]. COPD is a serious long-term condition that progressively restricts airflow from the lungs and imposes a significant burden on patients' daily lives [23]. According to a projection from the world health organization, by 2030, COPD will be the fourth most common cause of death [18], imposing a high socioeconomic burden worldwide [17]. More than 3 hundred thousand Norwegians are suffering from COPD [22].

Patients with COPD often suffer from exacerbations, meaning a sustained worsening of symptoms which leads to hospitalization. There is growing evidence that exacerbation incidents lead to lung function decay, reduced quality of life, more hospital re-admissions monitoring, and increased health care expenditure [21]. An approach to predicting COPD exacerbations prior to their occurrence is a natural step, but this is far from trivial.

There are many attempts at both medical diagnosis and predicting a patients course of illness using machine learning in general and deep learning models in particular. The success of applying deep learning tends to correlate with the size of the training dataset, as these models need large datasets to successfully learn the needed patterns from patients medical history. This is an undesirable property as medical observations are often of limited size.

In our study we explore whether deep learning models, and particularly Long-Short Term Memory (LSTM) can be used to predict exacerbation/condition of COPD patients even when the dataset is limited such as is typical in the case of home monitoring situations. Hence, we want to predict the COPD patients with acute exacerbation in advance which can lead to better patient care and decision making, and hence better quality of life, and a reduction in costs.

The rest of paper is organized as follows. In Sect. 2, we briefly discuss some of the previous work in the field of applying ANN for COPD. In Sect. 3, the research methodology that we have applied is explained. Section 4 gives the description of our considered dataset and the data pre-processing and analysis process. In Sect. 5, we present the experimental results. Finally, Sect. 6 concludes our paper with future developments.

2   Related Work

Researchers have proposed and applied various deep learning methods to different kinds of diseases for prediction and classification [5–7,13,16,19,20,27]. However, in this paper, we present the state-of-the-art related to applying deep learning for the COPD.

In the study [1], the authors worked on prediction of COPD acute exacerbations by using a feed forward multi-layer neural network (FFNN). The mean square error was computed on three assigned testing data sets, the scores were 0.1312, 0.1380, and 0.1041.

In [2], the authors worked on predicting which patients were at high risk for multiple COPD exacerbations and hospital re-admission within a single year. To do so, they used random forest (RF) algorithm. In their study, mean area-under-curve (AUC) statistics, sensitivity, specificity, and negative/positive predictive values (NPV, PPV) were calculated. The scores of mean AUC was 0.72, sensitivity was 0.75, specificity was 0.56, PPV was 0.7 and NPV was 0.63.

In [3], the authors worked on classifying the COPD patients' lung function based on fuzzy rules and an ANN. To achieve the classification, the authors considered 285 COPD patients with 92% correct classification.

In [4], the authors worked on recognizing and classifying the COPD patients and specifically whether they were suffering from central and peripheral airways obstructions. For classification, the authors used 131 patients' dataset and applied feed-forward ANN. The study results show that the authors achieved 98.47% correct classification rate. When a new set of unseen data was used, they acquired 61.53% classification accuracy.

In [9], the authors have worked on realizing COPD diagnosis by using multi-layer neural networks (MLNN). For this purpose, two different MLNN structures were used: a one hidden layer MLNN, and a two hidden layers MLNN. Back propagation (BP) with momentum and Levenberg-Marquardt (LM) algorithms were used for the training of the neural networks. The authors achieved 93.14% when they used BP algorithm with one hidden layer and 94.46% when they trained the FFNN model with LM algorithm and two hidden layers.

In [10], the authors worked on evaluating the feasibility respiratory sensors, computerized analyses of respiratory sounds, and tele-health framework for early detection of the acute exacerbation COPD by using machine learning algorithms called principal component analysis (PCA) and support vector machine (SVM) classifier. The study results showed a predictive capacity for exacerbations of 75.8% with an average of 5 ± 1.9 days in advance of medical attention.

Whereas, in [12], the authors worked on classifying the COPD four disease levels (mild, moderate, severe; very severe) with Artificial Neural Networks (ANN) with 507 patients dataset. They used 2 hidden layers and 5 layered cross-validation method. The results found the ANN model accuracy to be 99% and the computed MSE and MAE values 0.00996, and 0.02478 respectively.

The authors in study [23] worked on early detection of exacerbations and subsequent triage of COPD patients. Two algorithms called gradient-boosted decision tree (GBDT) and the logistic regression (LR) were used in identifying exacerbations and predicting the consensus triage. The authors achieved 88.1% accuracy for triage classification when GBDT was used and 89.1% accuracy when LR algorithm was used. However, 97.0% accuracy was achieved for early detection of exacerbations when both the gradient-boosted decision tree and the logistic regression algorithms were used.[1]

In [26], the authors built and compared three predictive machine learning models such as regularized logistic regression (LASSO), gradient boosting machine (GB), and multi-layer perceptron (MLP) approach to predict 30-day hospital re-admissions of COPD patients and to compute the prediction performance accuracy. They used AUC as the measure of prediction performance. They achieved model accuracy of 0.700 with LASSO, with GB, they achieved model accuracy of 0.706, and with MLP, they achieved model accuracy of 0.705.

---

[1] This study was carried out on a different large dataset than our study and is therefore not comparable.

Although various scholars have worked on predicting the acute COPD exacerbations with various machine learning algorithms, none of the above mentioned studies examined the applicability of recurrent neural networks, including LSTM, for predicting the acute COPD exacerbations based on time intervals. Therefore, in this research, we will explore the usefulness of the recurrent neural networks for predicting the acute COPD exacerbations.

3   Research Approach

Our approach aims at predicting the COPD exacerbations prior to their occurrence from COPD data. This should ideally help medical nurses and doctors in decision making. In addition to predicting future COPD exacerbations, we have classified our data by using feed-forward neural network (FFNN). The developed FFNN consists of an input layer (5 neurons), one hidden layer with 8 neurons and an output layer with 3 neurons. We used the "adam" optimization algorithm [15], the rectified linear unit (ReLU) function as activation function, and one dropout layer is included to prevent over-fitting.[2]

To predict the COPD exacerbations prior to their occurrence, we train the LSTM architecture on COPD data. The used an LSTM architecture that contains one hidden layer containing 64 LSTM blocks presented. The input of the network is a vector representing values such as the patient IDs, the date when the patients took the measurements, the measured value of the SpO2 of the blood, the measured value of the pulse, the result of a questionnaire overview, the SpO2 reference value of each individual patient, and the pulse reference value of each individual patient. The output node of the network Ot is a softmax distribution over the labels [14]. The softmax function produces an output in the [0, 1] interval that represents the probability of the triage given the input of the node. For a training set $(x1, y1), (x2, y2), .., (xn, yn)$ (here x represents the feature and y represents the labelled information), the softmax distribution is defined by Eq. 2.

$$p(y^i = k|x^i) = \frac{exp(\theta_i x^i)}{\sum_{j=1}^{K} exp(\theta_j x^i)} \tag{1}$$

where $x^i \in IN^K$ is vector coding of the input of the softmax (note that in our case the input of the softmax is the output of the hidden layers of the LSTM architecture), $y^i = 1, 3, 5$ is the index of the output vector, K is the number of possible categories. θ^i is the parameter of the softmax to be determined during the training phase to minimize the loss function in Eq. 2. The loss function represents the sum of the negative categorical crossentropy of y_i knowing x^i. By minimizing the loss function, the probabilities that the correct feature is predicted approaches one.

$$J(\theta) = -\frac{1}{n} \sum_{i}^{n} y_i log p(\hat{y_i}) \tag{2}$$

---

[2] The experimental setup is available in https://github.com/cair/TELMA-Project.

4   Data Processing

The models for classification and prediction of patients with COPD exacerbations used the COPD dataset described in the following sub-sections.

4.1   The Data

The data that has been used for predicting COPD exacerbations was extracted in the EU-funded-project "United4Health" [24,25] during a real home monitoring session. The dataset contains daily measurements of symptom-specific questionnaire, pulse, peripheral capillary oxygen saturation (SpO2) of the blood, and results of automatically generated health status overview of all monitored patients who are at home. An automatically generated health status overview (triage), is generated based on a medical rule set and colour coded as green, yellow, and red. For statistical purposes, the triage was labeled as 1, 3, and 5. Here green (1) means, the patient was stable, yellow (3) means the patient had a notable deterioration, while red (5) means most urgent need for follow-up. In total, the dataset is of 96 patients and over 7300 rows for a period of two years. More details about the dataset can be found in [11].

4.2   Data Preprocessing and Analysis

To train the LSTM model, typically a large amount of data is needed to achieve good performance. However, such large datasets are often not available. Therefore, a data augmentation technique has been built to boost the performance of the proposed model, this is a powerful model using very little data. Here, data-augmentation mean augmenting the training dataset via domain-specific transformations such as random rotation, and shifts [28]. In our research we also used data-augmentation technique to artificially increase the size of the training set, as the taken COPD dataset is relatively small for the deep learning to achieve good performance of the proposed models.[3]

The considered COPD data include a lot of 1 and 3, but not 5. Hence, first we calculated the number of data entries with class 5 that was needed to balance out the dataset. After this, we proceeded by looping through the data and for every entry containing class 5 the necessary amount of copies were created. Each copy was then modified by adding gaussian noise before being introduced into the training data. After adding Gaussian noise, we divided the set into 2 separate data sets, 50% for training and 50% for testing. After splitting training and testing data, we further split the training data into 10 parts to check how much data the training dataset needed to get good model accuracy for the prediction, so that cross-validation was still kept. This was to ensure that no variant of the same data was used in both training and validation, even noisy variants.

---

[3] Additional experiments without data-augmentation was carried out, but are for reasons of space deliberately left out of the paper. Without data-augmentation in the training set, the classifier performs badly.

To do so, we ran the model each time by increasing the training data 1 part (i.e., 10%) each time until 10th part (i.e., 100%). For one-day or 2-days or 3-days ahead prediction, we specified and processed the data with a time window of past 5 days such as $t, t-1, t-2, t-3$, and $t-4$. Based on this time window, the model predicts next 3 days as $t, t+1, t+2$.

5 Experimental Results

Our experiments are divided into two parts classification and prediction. Before showing prediction of acute exacerbations of patients with COPD, in the first set of experiments we show that we are able to classify exacerbations. This can be interpreted as a zero day predictions and we show that the FFNN is able to reproduce the medical expertise. In the second set of experiments, we show that we can foresee upcoming exacerbations days in advance. The models are built using the Keras deep learning library in Python.

5.1 Classification

Two approaches are implemented a key part of a decision support tool for the medical expertise on categorizing patients into one of the three categories: 1, 3, and 5. The first approach is a standard FFNN model and the output of the FFNN network is the exacerbation class, and the second is an LSTM model for prediction set to predict 0 days ahead. See Sect. 3 for details on the models.

For the FFNN model, all the given COPD data available was used for training and classifies which class the COPD patients are belong to. For the FFNN model, the COPD data was trained for 150 epochs with a batch size of 3, achieving an 92.86% accuracy. This results show that the FFNN trained with the measurements provided by the COPD patients is capable of classifying the patients who belong to which category with a higher accuracy.

For the LSTM model, the COPD data was trained for 50 epochs with a batch size of 10. We run the experiment for 50 times. When the model was trained, 70% accuracy was obtained for classifying the COPD may be belong to either class 1 or 3 or 5 and predicting patients with acute exacerbation for the same day. It means that the obtained result show us that it can be able to do the same as the medical people would do in the same situation.

5.2 Prediction

With predictions we mean predicting medical exacerbations at least one day prior to their occurrence. The prediction model is the same as LSTM classification (0 day prediction), and for all the presented below experiments, the COPD data fed to the LSTM model and trained for 50 epochs with a batch size of 10. We ran the experiment for 50 times, performed several experiments and realized the prediction accuracy is very erratic. A natural assumption to make is that this is caused by the small limited dataset. Part of our experiments is centred around ways to improve this, including carefully removing records from the data.

Experiment 1 - All Data: In our first experiment, we kept all the records of all the three classes (AD) in the taken dataset and fed to the LSTM model to predict one-day-ahead. The model reached a prediction accuracy of 82.5% (see Fig. 1) with 0.50 testing loss. The highest accuracy is achieved with 80% of the data for training. The performance of the LSTM model with all data is presented in confusion matrix (see matrix in Table 1). The matrix shows that 315, 202 and 2 samples are correctly predicted as class 1, 3, and 5. respectively. An hypothesis is that increasing the training data, increases the quality of the model. If this trend is present with our data, it is natural to assume the quality would keep increasing beyond the data size we have available. We cannot prove anything beyond our data, but we can claim that data size matters. Our experiments show that this is partly true as the best accuracy was with 80% of the data.

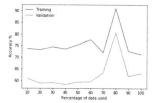

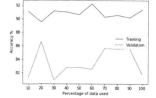

Fig. 1. The LSTM model accuracy when all the three classes present.

Fig. 2. The LSTM model accuracy when class 3 is removed.

Further, the maximum likelihood is also calculated from the output vector of the neural net to show the percentage wise likelihood of a prediction per class. The likelihood of predicting a true class 1 as a class 1 patient is 56.45%, 35.59% as a class 3 patient and 7.94% as a class 5 patient. When the true class is, the likelihood of classifying it as class 1, 3 and 5 is 40.40%, 49.39% and 10.19% respectively. Lastly, when the correct class is a category 5 patient, the likelihood for 1, 3 and 5 is 46.42%, 42.55% and 11.01% respectively. Note that all our models over-train on the data, which is not surprising given the small data set. This suggests that independent of how we split the data there will always patterns in the training examples not in the validation data.

Experiment 2 - Removing Class 3 (yellow): In our second experiment, we removed all the records of class 3 (C3) in the taken dataset and fed to the LSTM model to predict one-day-ahead. The motivation for removing class 3 is two-fold, and the primary reason is that, by close examination of the taken COPD data, several of the class 3 data were wrongly classified and arguably reduces the prediction accuracy of the model. This is not surprising since class 3 patients lie in between class 1 and 5 and the patient data therefore presumable overlaps with both. It begs the question, what would the accuracy be without the hard class 3 cases. The authors are of course aware that simply removing data which is hard to classify in no way resembles best practice. However, the secondary reason is that the medical personnel is interested in the extreme cases, which is class 5 cases. If a patient will be in class 5 in few days, measures are needed. Hence, the overall motivation of this paper in these situations is that the LSTM model is able to predict, and by removing category 3 patients, we get a better understanding of possibilities and limitations of the model.

Table 1. Confusion matrix including all

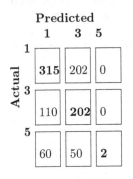

Table 2. Confusion matrix excluding 3

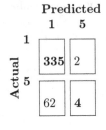

Table 3. Confusion matrix for SVM

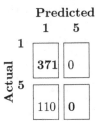

The model achieves a prediction accuracy of 84.12% (see Fig. 2) with 0.45 testing loss. The highest accuracy is achieved with 70% of the data for training. The performance of the LSTM model excluding class 3 is presented in confusion matrix (see matrix in Table 2). The matrix shows that 335 samples are correctly predicted as class 1, and 4 as class 5. Further, a maximum likelihood is calculated to show the percentage wise likelihood of a prediction per class. 84.73% is the likelihood of correctly predicted as a class 1 patient as class 1, and consequently as class 5 with a 15.26% likelihood. Moreover, for the class 5, 22.5% is the likelihood of a correct classification, with and likelihood of 77.47% as 1. We observe that predicting class 1 patients is still much easier than class 5 patients.

Experiment 3 - Three Days Ahead Prediction: In our third experiment, we removed all the records of class 3 (yellow) in the taken dataset and fed to the LSTM model to predict next three days ahead. For one-day ahead prediction, the model achieves a prediction accuracy of 84.12% (see Table 4) with 0.45 testing loss. The highest accuracy is achieved with 70% of the data for training. For second-day ahead prediction, the model achieves a prediction accuracy of 88% (see Fig. 3) with 0.38 testing loss. The highest accuracy is achieved with 50% of the data for training. For third-day ahead prediction, the model achieves a prediction accuracy of 88% (see Fig. 3) with 0.345 testing loss. The highest accuracy is achieved with 60% of the data for training. From the results, we observe that the prediction model for one-day ahead is more stable.

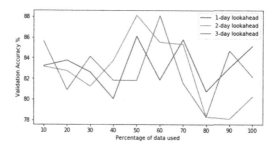

Fig. 3. Predicting next three days with LSTM model.

Table 4. Comparing our results with existing results.

Literature	Method	Classification	Prediction (detection)
[1]	FFMLNN	-	MSEs = 0.1312, 0.1380, and 0.1041
[2]	RF algorithm	AUC = 0.75	-
[3]	FFMLNN	92%	-
[4]	FFMLNN	98.47%	-
[9]	BP	93.14%	-
	LM algorithm	94.46%	-
[10]	PCA	-	
	SVM	-	75.8%
[12]	MLP	99%	-
[23]	GBDT	88.1%	97%
	LR	89.1%	97%
[26]	RLR(LASSO)	-	AUC(LASSO) = 0.700
	GB	-	AUC(GB) = 0.706
	MLP	-	AUC(MLP)= 0.705
Our work	**FFNN**	**92.86%**	-
	LSTM	**70%**	**AD = 82.5%, C3 = 84.12**
	SVM	**59.61%**	**AD = 59.65%, C3 = 77.13%**

6   Conclusions and Future Work

This work explores the potential of using several types of artificial neural networks for classifications of COPD patients' category and predictions of acute exacerbations of patients with COPD using a real data set from a realistic home monitoring situations. More specifically, we have applied FFNN for classification of COPD patients and LSTM for predictions of acute exacerbations of patients with COPD prior to their occurrence as a way to investigate the applicability of distance monitoring of patients with chronic diseases.

In order to classify the COPD patients' category, we used FFNN and LSTM machine learning approaches and obtained 92.86% and 70% model accuracy respectively. However, for predicting 3 days ahead, we have done several experiments. The experiment results show that for 1 day ahead prediction, 84.12% accuracy is obtained. Whereas, we achieved 88% prediction accuracy for 2 days ahead, and 88% prediction accuracy for 3-days ahead. For comparison purpose, we have also applied SVM with a RBF kernel which reached an accuracy of 59.61% model accuracy is obtained for classification and reached 59.65% accuracy for predicting one-day ahead when all data is used for training, and achieved 77.13% accuracy predicting one-day ahead when class 3 is removed.

When class 3 is removed, the calculated confusion matrix for SVM model prediction shows that 371 samples are correctly predicted as class 1, and 0 as class 5 (see Table 3). With this, it can be seen that SVM is not able to learn sufficiently from the data. Further, the class 5 category is also not able to be predicted by using SVM. Moreover, if we compare our results with the existing literature (see Table 4), for 1-day ahead prediction, our results show higher accuracy than the [10] when both LSTM and SVM are used, but compared with [23], they have larger and different dataset, making the results incomparable.

As future work, we plan to carry out the following research: classification the medical condition and prediction other chronic diseases such as heart congestion, psychiatric diseases and even comorbidity patients. Most importantly, this work is planned to be used by medical personnel for decision making.

Acknowledgements. This work is part of an ongoing the Norwegian Research council project TELMA (Telemedicine in Agder) which develops a common solution for distance monitoring of patients with chronic diseases and comorbidity at Agder region to provide good health care with less use of health resources. We would like to thank Martin Wulf Gerdes and Prof. Rune Werner Fensli for sharing the COPD data.

References

1. Alharbey, R.: Predictive analytics dashboard for monitoring patients in advanced stages of COPD. In: IEEE 49th Hawaii International Conference on System Sciences (HICSS), pp. 3455–3461 (2016)
2. Amalakuhan, B., Kiljanek, L., Parvathaneni, A., Hester, M., Cheriyath, P., Fischman, D.: A prediction model for COPD readmissions: catching up, catching our breath, and improving a national problem. J. Community Hosp. Intern. Med. Perspect. **2**(1), 99–115 (2012)
3. Badnjevic, A., Cifrek, M., Koruga, D.: Classification of chronic obstructive pulmonary disease (COPD) using integrated software suite. In: XIII Mediterranean Conference on Medical and Biological Engineering and Computing, pp. 911–914 (2014)
4. Barúa, M., Nazeran, H., Nava, P., Granda, V., Diong, B.: Classification of pulmonary diseases based on impulse oscillometric measurements of lung function using neural networks. In: IEEE 26th Annual International Conference of the Engineering in Medicine and Biology Society, pp. 3848–3851 (2004)

5. Che, C., Xiao, C., Liang, J., Jin, B., Zho, J., Wang, F.: An RNN architecture with dynamic temporal matching for personalized predictions of parkinson's disease. In: The International Conference on Data Mining, pp. 198–206. SIAM (2017)
6. Choi, E., Schuetz, A., Stewart, W.F., Sun, J.: Using recurrent neural network models for early detection of heart failure onset. J. Am. Med. Inform. Assoc. **24**(2), 361–370 (2016)
7. Cruz, J.A., Wishart, D.S.: Applications of machine learning in cancer prediction and prognosis. Int. J. Cancer Inform. **2**(1), 1–19 (2006)
8. Er, O., Sertkaya, C., Temurtas, F., Tanrikulu, A.C.: A comparative study on chronic obstructive pulmonary and pneumonia diseases diagnosis using neural networks and artificial immune system. J. Med. Syst. **33**(6), 485–492 (2009)
9. Er, O., Temurtas, F.: A study on chronic obstructive pulmonary disease diagnosis using multilayer neural networks. J. Med. Syst. **32**(5), 429–432 (2008)
10. Fernandez-Granero, M.A., Sanchez-Morillo, D., Leon-Jimenez, A.: Computerised analysis of telemonitored respiratory sounds for predicting acute exacerbations of COPD. Int. J. Sens. **15**(10), 26978–26996 (2015)
11. Gerdes, M., Gallefoss, F., Fensli, R.W.: The EU project "United4Health": results and experiences from automatic health status assessment in a Norwegian telemedicine trial system. J. Telemed. Telecare **23**, 1–8 (2017)
12. Işık, Ü., Güven, A., Büyükoğlan, H.: Chronic obstructive pulmonary disease classification with artificial neural networks. In: IEEE National Conference on Medical Technologies (TIPTEKNO), pp. 1–4 (2015)
13. Kaur, G., Chhabra, A.: Improved J48 classification algorithm for the prediction of diabetes. Int. J. Comput. Appl. **98**(22), 1–5 (2014)
14. Kawakami, K.: Supervised sequence labelling with recurrent neural networks. Ph.D. thesis, Technical University of Munich (2008)
15. Kingma, D.P., Ba, J.: Adam: a method for stochastic optimization. In: International Conference on Learning Representations, pp. 1–15 (2014)
16. Latif, S., Usman, M., Rana, J.Q.R.: Abnormal heartbeat detection using recurrent neural networks. J. Comput. Vis. Pattern Recogn. **3**, 1–8 (2018)
17. Mannino, D.M., Buist, A.S.: Global burden of COPD: risk factors, prevalence, and future trends. The Lancet **370**(9589), 765–773 (2007)
18. Mathers, C.D., Loncar, D.: Projections of global mortality and burden of disease from 2002 to 2030. J. PLoS Med. **3**(11), 1–20 (2006)
19. Miotto, R., Li, L., Kidd, B.A., Dudley, J.T.: Deep patient: an unsupervised representation to predict the future of patients from the electronic health records. Nat. Sci. Rep. **6**, 1–10 (2016)
20. Moradi, E., Pepe, A., Gaser, C., Huttunen, H., Tohka, J., Alzheimer's Disease Neuroimaging Initiative, et al.: Machine learning framework for early MRI-based Alzheimer's conversion prediction in MCI subjects. Int. J. Neuroimage **104**, 398–412 (2015)
21. Shah, S.A., Velardo, C., Farmer, A., Tarassenko, L.: Exacerbations in chronic obstructive pulmonary disease: identification and prediction using a digital health system. J. Med. Internet Res. **19**(3), 1–9 (2017)
22. Sintef: most COPD patients in southern and eastern Norway (2014). https://www.sintef.no/en/latest-news/most-copd-patients-in-southern-and-eastern-norway/
23. Swaminathan, S., et al.: A machine learning approach to triaging patients with chronic obstructive pulmonary disease. J. PloS One **12**(11), 1–21 (2017)
24. United4Health: FP7 EU project: United4health (norwegian sub-project) (2015). http://www.united4health.no/information-in-english/

25. United4Health: FP7 EU project: Transforming the patient experience with tele-health in europe (2017). http://www.united4-health.eu/
26. Yang, C., Delcher, C., Shenkman, E., Ranka, S.: Predicting 30-day all-cause re-admissions from hospital inpatient discharge data. In: IEEE 18th International Conference on e-Health Networking, Applications and Services (Healthcom), pp. 1–6. IEEE (2016)
27. Yu, W., Liu, T., Valdez, R., Gwinn, M., Khoury, M.J.: Application of support vector machine modeling for prediction of common diseases: the case of diabetes and pre-diabetes. BMC Med. Inform. Decis. Mak. **10**(1), 1–16 (2010)
28. Zhang, C., Bengio, S., Hardt, M., Recht, B., Vinyals, O.: Understanding deep learning requires rethinking generalization. In: International Conference on Learning Representations, pp. 1–15 (2016)

Probabilistic Word Association for Dialogue Act Classification with Recurrent Neural Networks

Nathan Duran$^{(\boxtimes)}$ and Steve Battle

University of the West of England, Coldharbour Ln, Bristol BS16 1QY, UK
{nathan.duran,steve.battle}@uwe.ac.uk

Abstract. The identification of Dialogue Act's (DA) is an important aspect in determining the meaning of an utterance for many applications that require natural language understanding, and recent work using recurrent neural networks (RNN) has shown promising results when applied to the DA classification problem. This work presents a novel probabilistic method of utterance representation and describes a RNN sentence model for out-of-context DA Classification. The utterance representations are generated from keywords selected for their frequency association with certain DA's. The proposed probabilistic representations are applied to the Switchboard DA corpus and performance is compared with pre-trained word embeddings using the same baseline RNN model. The results indicate that the probabilistic method achieves 75.48% overall accuracy and an improvement over the word embedding representations of 1.8%. This demonstrates the potential utility of using statistical utterance representations, that are able to capture word-DA relationships, for the purpose of DA classification.

Keywords: Dialogue acts · Neural networks · Probabilistic

1 Introduction

The notion of a Dialogue Act (DA) originated from John Austin's 'illocutionary act' theory [1] and was later developed by John Searle [25], as a method of defining the semantic content and communicative function of a single utterance of dialogue. The utility of DA's as a set of labels for a semantic interpretation of a given utterance has led to their use in many applications requiring Natural Language Understanding (NLU). In dialogue management systems they have been used as a representation of user and system dialogue turns [5] or as a set of possible system actions [3]. For spoken language translation Kumar et al. [13] utilized the contextual information provided by DA's to improve accuracy in phrase based statistical speech translation.

To facilitate their use in such systems, utterances must first be assigned a single DA label, sometimes called short text classification. Previously, many different approaches have been applied to the DA classification problem, including

© Springer Nature Switzerland AG 2018
E. Pimenidis and C. Jayne (Eds.): EANN 2018, CCIS 893, pp. 229–239, 2018.
https://doi.org/10.1007/978-3-319-98204-5_19

Support Vector Machines (SVM) and Hidden Markov Models (HMM) [27], n-grams [17] and Bayesian networks [11]. More recently Artificial Neural Network (ANN) based approaches have led to increased performance, particularly Convolutional Neural Networks (CNN) [10] and Recurrent Neural Networks (RNN) [8,14,22]. Many of these approaches consider the DA classification task on both a sentential and discourse level. The sentence level is concerned with how the order and meaning of words compose to form the meaning of a sentence [15]. Similarly, on a discourse level, the order and meaning of sentences compose to form the meaning of sequences in dialogue [24]. Certainly, a given utterance and its associated DA is often directly influenced by the nature of the preceding utterances and the current context of the dialogue. For example, 'okay' could be an acknowledgement of understanding or an agreement to a request, the intention is dependent on the utterance it is responding to. This view has led much of the neural network based research to model the semantic content of a sentence, in conjunction with some other contextual information, such as previous utterance or DA sequences, or a change in speaker turn, to predict the appropriate DA for the current utterance [10,14]. Including such historical and contextual information has been shown to improve classification accuracy [16], and likely must be considered for any sophisticated classification model. However, motivated to examine the importance of different lexical and syntactic features, and their contribution towards DA classification, in this work each utterance is considered individually and out-of-context, that is, without any other contextual or historical information. Further, Cerisara et al. [2] determined that the use of traditional word embeddings, such as GloVe [23] and Word2vec [20], have a limited impact on the DA classification task and therefore, this work explores an alternative approach for utterance representations.

This work presents a simple, yet effective, probabilistic method of utterance representation and DA classification using RNN architectures. The utterance representations are generated from the probability distribution over all DA's for each word in the utterance. Intuitively, each word is represented by a vector of the probabilities that it is associated with each DA, and an utterance is then a matrix formed from the vector representations of the words it contains. This representation method was inspired by the intuition that certain keywords may be associated with certain DA types and therefore act as indicators for the DA of the utterances that contain them [26]. A Long-Short Term Memory (LSTM) network based model is then used to classify the utterances according to their associated DA types. A description of the model and utterance representations can be found in Sect. 3. Experimentally this method is applied to the DA classification task using the Switchboard Dialogue Act (SwDA) corpus (Sect. 4) and yields results comparable with more sophisticated approaches that also consider utterance or dialogue context information. Performance for representations generated from different word frequencies is compared to traditional word vector representations using the same LSTM model in Sect. 5. The following section is a discussion of neural network architectures for DA classification and cue-phrase and n-gram approaches with similar motivations to this probabilistic word representation method.

2   Related Work

The ability of RNN to model long term dependencies in sequential data has led to their widespread use in many Natural Language Processing (NLP) tasks [21], and recently both LSTM and CNN have been applied to the DA classification problem with great success. These approaches commonly employ a combination of a sentence model and a discourse or context model. The sentence model acts at the utterance level and encodes sentences, often from word embeddings, into a fixed length vector representation. The encoded utterances are then used as input to a discourse level model that incorporates some other contextual information and classifies the current utterance. For example, Lee and Dernoncourt [14] experiment with both LSTM and CNN sentence models for encoding short text representations, followed by various ANN architectures for classifying the encoded utterances based on the current, and up to two of the previous, utterances. To try and capture interactions between speakers, Kalchbrenner and Blunsom [10] used a Hierarchical CNN sentence model in conjunction with a RNN discourse model and condition the recurrent and output weights on the current speaker. Liu et al. [16] examine several different CNN and LSTM based architectures that incorporate different context information such as speaker change and dialogue history. Their work shows that including context information consistently yielded improvements over their baseline system.

These approaches all use pre-trained word embeddings to construct utterance representations for the input to a sentence model. While word embeddings carry some semantic similarity information useful for many language classification tasks [18], they do not convey any relational information between the words in an utterance and its associated DA. The work of Cerisara et al. [2] showed that pre-trained embeddings did not help the DA classification task and this is likely due to the word vector training corpora commonly being non-conversational. In an effort to incorporate word-DA relationships into a similar model as those already discussed, Papalampidi, Iosif and Potamianos [22] explored the use of keywords that are representative of DA's. First a set of keywords was constructed based on word frequency and saliency with respect to each DA. The keywords were then used to add a weighting value to pre-trained word embeddings for input to an LSTM sentence model. A two-layer ANN then classified utterances based on the current and preceding two utterances in a similar fashion as [14].

The notion that certain words or phrases can act as indicators for utterance DA labels is more often explored via probabilistic methods. Garner et al. [4] described a theory of word frequencies for dialogue move recognition and concluded that better performance could be achieved using a more involved n-gram model, such as that applied by Louwerse and Crossley [17]. Webb and Hepple [28], selected a set of 'cue phrases' based on the probability of an n-gram occurring within a given DA and keeping only those with the 'predictivity' values over a certain threshold. Utterance classification is then performed by identifying the cue phrases it contains and assigning a DA label based on the phrase with the highest predictivity value. The probabilistic methodology in this paper combines aspects of the previously described approaches, sing an LSTM sentence model with utterance representations generated from probabilistic word-DA relations.

3   Model

3.1   LSTM Sentence Model

The sentence model is similar to those used by Papalampidi, Iosif and Potamianos [22], and also Khanpour, Guntakandla and Nielsen [12], and is based on a standard LSTM network as described by Hochreiter and Schmidhuber [7]. A given utterance that contains n words, is converted into a sequence of $n \times m$ dimensional vectors $V_1, V_2, \ldots, V_n$. Where m is either the dimension of the word embeddings, or the number of DA in the case of the probabilistic representations (see Sect. 3.2). The lexical order of the words in the original sentence are represented as a sequence of successive time-steps in V. This sequence is given as input to the LSTM which produces an h dimensional vector at each time-step n, where h is the size of the LSTM hidden dimension. A pooling layer then combines the output vectors from each time-step $h_1, h_2, \ldots, h_n$ into a single vector representation s of the utterance. Finally, a single feed forward layer computes the probability distribution over all DA using the softmax activation function (Fig. 1).

Following the initialisation parameters proposed by Ji, Haffari and Eisenstein [8], with the exception of the bias (**b**) and output (**U**) matrices, all LSTM weights are initialised in the range $\pm\sqrt{6 \div (d_1 + d_2)}$ where d_1 and d_2 are the

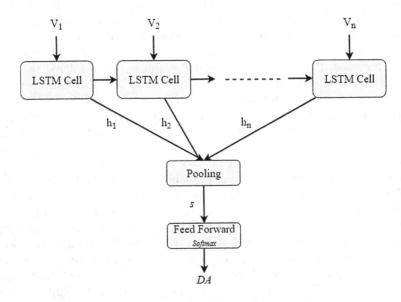

Fig. 1. LSTM sentence model

input and output dimensions respectively. U is initialised with random uniform distribution in the range ± 0.05 and $\mathbf{b}$ is initialised to 0. Optimisation is performed using the RMSProp algorithm with an initial learning rate $\alpha = 0.001$ and decay rate $\gamma = 0.001$.

3.2 Probabilistic Word Representations

The probabilistic word vector representations are simply generated by calculating a probability distribution over each of the DA's for each word that occurs at, or above, a certain frequency in the corpus vocabulary. First a set of n keywords is created keeping only those that occur at a frequency equal to or greater than a frequency threshold value (see Sect. 5.2). Using this set of keywords, a probability matrix X is created of size $n \times m$, where m is the number of DA used in the corpus. Thus, an element x_{ij} in X will be the probability that the ith word in n appears in an utterance that has the corresponding j^{th} DA tag in m. Each row in X is a probability distribution for the ith word in n for all DA's, and is then effectively a Probabilistic word embedding for creating the utterance representations as described in Sect. 3.1.

4   Experimental Dataset

The Switchboard Dialogue Act (SwDA) contains 1155 transcripts of $\sim$5-minute telephone conversations between two speakers that did not know each other and were provided a topic for discussion. Each of the $\sim$205,000 utterances is annotated with one of 42 DA tags using the Discourse Annotation and Markup System of Labelling (DAMSL) [9]. The transcripts were split into the same 1115 for training and 19 for testing as used by Stolcke et al. [26] and others [8,10]. The remaining 21 transcripts were used as a validation set and 300 were randomly selected from the training set for development purposes (Table 1).

Table 1. Datasets

Dataset	# of Transcripts	# of Utterances
Training	1115	192,768
Development	300	51,611
Test	19	4,088
Validation	21	3,196

The transcripts were pre-processed to remove the disfluency (breaks or irregularities), and other annotation symbols, in order to convert each utterance into a plain text sentence. Additionally, any utterances tagged as 'Non-Verbal', such as laughter or coughing, were removed from the transcript, as these do not contain any relevant lexical information. In this work therefore, 41 of the original 42 DAMSL tags were used, reducing the utterance count by $\sim$2% (Table 2).

Table 2. Most frequent Switchboard DA tags.

Dialogue Act	Tag	Example	Count	%
Statement-non-opinion	sd	*Me, I'm in the legal department.*	75,138	37%
Acknowledge (Backchannel)	b	*Uh-huh.*	38,233	19%
Statement-opinion	sv	*I think it's great*	26,422	13%
Abandoned or Turn-Exit	%-	*So, -*	15,545	7%
Agree/Accept	aa	*That's exactly it.*	11,123	5%
Appreciation	ba	*I can imagine.*	4,759	2%
Yes-No-Question	qy	*Do you have special training?*	4,726	2%
Yes answers	ny	*Yes.*	3,031	1%

5   Results and Discussion

5.1   Parameter Tuning

As previously stated, the same LSTM model is used for comparison between the traditional word embedding utterance representations and the Probabilistic representations. Parameters were first tuned using traditional word embeddings and then applied to the Probabilistic representations. Keeping the hyperparameters fixed, different word-to-vector techniques and dimensions are tested. Each parameter is then tuned one at a time to determine the optimum configuration. Both l2-regularization and decay rate were also tested, though were shown to only have a negative impact on performance, and therefore l2-regularization is not used and decay is fixed at $\gamma = 0.001$. For all parameter tuning and utterance representation testing the development set described in Sect. 4 is used. Results shown are averaged over 5 runs of 10 epochs.

Word Embeddings. Two pre-trained sets of word embeddings were tested, word2vec [19] trained on the Google News corpus and GloVe [23] trained on a Wikipedia corpus. In addition, a second word2vec embedding was trained on the SwDa corpus itself using the Gensim python package.[1] Dimensions in the range 100–300 were tested and hyperparameters were kept fixed with the values; dropout = 0.3, hidden dimension = 64, and max-pooling. Table 3 shows the best classification accuracy was achieved using word2vec trained on Google News with 300 dimensions and this is adopted for the rest of the experiments.

Dropout. Dropout is a regularisation method generally used to prevent overfitting in neural networks [6]. The results in Table 4 concur with others' findings [12,14,22] that a value of 0.3 is optimal.

---

[1] https://radimrehurek.com/gensim/.

Table 3. Word embeddings performance.

Word Embeddings	Embedding dimension	Test set accuracy %	Validation set accuracy %
GloVe wiki	100	70.62	74.22
	200	71.31	74.98
	300	70.45	73.68
Word2vec google news	100	69.41	72.97
	200	71.14	75.06
	300	**71.65**	**75.07**
Word2vec SwDa	100	68.93	72.58
	200	68.71	72.29
	300	69.18	72.81

LSTM Hidden Dimension. The LSTM hidden state corresponds to the dimensionality of the output vectors of an LSTM cell at each time-step $h_1, h_2, \ldots, h_n$ and therefore determines the dimension of the utterance representation s that is generated by the pooling layer. Table 4 shows minimal impact on performance provided the hidden dimension is close to the maximum sentence length ($\sim$107 words).

Pooling. Two different pooling mechanisms are tested. Max-pooling keeps the element-wise maximum of the h vectors output by the LSTM and mean-pooling averages the h vectors.

Table 4. Hyperparameter tuning performance.

Dropout	Hidden dimension	Pooling	Test set accuracy %	Validation set accuracy %
0.0	64	Max	70.91	74.45
0.1			71.36	74.82
0.2			71.58	74.82
0.3			71.62	75.17
0.4			71.26	74.97
0.3	128	Max	**72.32**	**75.66**
	256		72.01	75.51
0.3	128	Mean	68.51	72.09

5.2 Probabilistic Word Embeddings

Different word frequency thresholds (see Sect. 3.2) were tested to explore the performance impact on the probabilistic word representation vectors. The thresh-

olds are simply an indication of the number of times a given word occurs in the
SwDa corpus. Table 5 shows accuracy tends to decrease with larger thresholds.
This is likely due to data for the utterance representations becoming too sparse
as fewer words are included in the probability matrix. For example, a threshold
of 2 eliminates around half of the words in the vocabulary. The thresholds minimum value was kept at 2 to maximise the likelihood of words appearing in at
least one of the test datasets.

Table 5. Performance for different word frequency thresholds.

Word frequency	Test set accuracy %	Validation set accuracy %
2	**74.68**	**77.38**
4	74.14	77.1
6	73.66	75.7
8	73.01	76.35
10	72.69	76.04

Table 6 shows a subset of the probability matrix created using the method
described in Sect. 3.2 for the SwDa corpus. It can be seen that certain words correlate significantly with specific DA's. These are particularly useful features for
differentiating between DA that are otherwise semantically similar, for example,
Statement-opinion and *Statement-non-opinion*. However, certain DA's such as
Abandoned/Turn-Exit do not correlate strongly with any words.

Table 6. Example of word probabilities for the five most common DA's.

	Statement non-opinion	Acknowledge (Backchannel)	Statement opinion	Abandoned turn-exit	Agree accept
My	86.67	0.03	4.88	1.52	0.08
Yeah	0.08	71.49	0.07	1.54	16.68
Should	26.22	0.00	59.19	0.40	0.40
Um	36.94	15.56	9.14	20.30	0.78
True	5.85	0.30	14.67	0.10	62.83

5.3   Results

Evaluation of the probabilistic and pre-trained word embedding representations
was performed on the full training set and results shown are an average over
10 runs for the test dataset. Table 7 shows the highest classification accuracies
achieved on the test dataset for both word representation methods using the
RNN model. The proposed model trained with word embeddings resulted in a

similar accuracy (73.68%) as the sentence model in the work of Papalampidi et al. [22]. The model trained on the Bayes representation shows an improvement of 1.8% over the word embeddings baseline. Though direct comparisons are difficult, due to differences in pre-possessing, models and other methodology, Table 7 also shows the probabilistic model is comparable to methods from the literature where context information was also used.

Table 7. Performance of the RNN model and other methods from the literature.

Model	Classification accuracy %
Sentence level	
Proposed LSTM - Probabilistic	**75.48**
Proposed LSTM - Word Embeddings	73.68
Sentence (Papalampidi et al. 2017)	73.8
Sentence and discourse level	
Sentence and Discourse (Papalampidi et al. 2017)	75.6
LSTM (Lee and Dernoncourt 2016)	69.6
CNN (Lee and Dernoncourt 2016)	73.1
RCNN (Kalchbrenner and Blunsom 2013)	73.9
DRLM-joint training (Ji et al. 2016)	74.0
DRLM-conditional training (Ji et al. 2016)	**77.0**
Inter-annotator Agreement (Stolcke et al. 2000)	84.0
Majority DA baseline	32.2

6 Conclusion

This work has presented a novel probabilistic approach to utterance representation, and an LSTM sentence model, for the purpose of DA classification. When applied to the SwDA corpus in an out-of-context fashion the probabilistic representations improve DA classification accuracy by 1.8% when compared to traditional word embeddings. Further, the overall highest classification accuracy achieved is competitive with approaches from the literature using more sophisticated classifier models that also consider contextual information, and improves on previously published results that only use a sentence level model. These findings also concur with previous work [2] to show that the traditional word embedding approach for utterance representation may not improve accuracy for the DA classification task. This highlights the need to find alternative representation methods for DA classification, such as the proposed probabilistic keyword-DA relationships.

Regarding future work, it would be beneficial to determine whether using probabilistic representations yields similar improvements in accuracy when using additional contextual information, and more sophisticated discourse and sentence models, which have been shown to improve results [14]. Additionally, it may be valuable to determine the portability of the approach by applying keywords gathered from one corpus to DA classification on another distinct corpus [29]. This would help to determine if certain keywords are able to generalise to new corpora, and perhaps reduce the amount of training data required.

References

1. Austin, J.L.: How To Do Things With Words. Oxford University Press, London (1962)
2. Cerisara, C., Král, P., Lenc, L.: On the effects of using word2vec representations in neural networks for dialogue act recognition. Comput. Speech Lang. **47**(July), 175–193 (2017). https://doi.org/10.1016/j.csl.2017.07.009
3. Cuayáhuitl, H., Yu, S., Williamson, A., Carse, J.: Deep reinforcement learning for multi-domain dialogue systems. In: NIPS Workshop Deep Reinforcement Learning, pp. 1–9. Barcelona (2016)
4. Garner, P.N., Browning, S.R., Moore, R.K., Russell, M.J.: A theory of word frequencies and its application to dialogue move recognition. In: 4th International Conference on Spoken Language Processing, (ICSLP 1996). ISCA, Philadelphia, PA (1996). https://doi.org/10.1109/ICSLP.1996.607999
5. Griol, D., Hurtado, L., Segarra, E., Sanchis, E.: A statistical approach to spoken dialog systems design and evaluation. Speech Commun. **50**(8–9), 666–682 (2008). https://doi.org/10.1016/j.specom.2008.04.001
6. Hinton, G.E., Srivastava, N., Krizhevsky, A., Sutskever, I., Salakhutdinov, R.R.: Improving neural networks by preventing co-adaptation of feature detectors. arXiv (2012). https://doi.org/arXiv:1207.0580
7. Hochreiter, S., Schmidhuber, J.: Long short-term memory. Neural Comput. **9**(8), 1735–1780 (1997)
8. Ji, Y., Haffari, G., Eisenstein, J.: A latent variable recurrent neural network for discourse relation language models. In: Proceedings of the NAACL-HLT 2016, pp. 332–342. ACL, San Diego (2016)
9. Jurafsky, D., Shriberg, E., Biasca, D.: Switchboard SWBD-DAMSL Shallow-Discourse-Function Annotation Coders Manual. Technical report (1997)
10. Kalchbrenner, N., Blunsom, P.: Recurrent convolutional neural networks for discourse compositionality. In: Proceedings of the Workshop on Continuous Vector Space Models and their Compositionality, pp. 119–126. ACL, Sofia, Bulgaria (2013). https://doi.org/10.1109/ICCV.2015.221
11. Keizer, S.: A bayesian approach to dialogue act classification. In: BI-DIALOG 2001 Proceedings of the 5th Workshop on Formal Semantics and Pragmatics Dialogue, pp. 210–218 (2001)
12. Khanpour, H., Guntakandla, N., Nielsen, R.: Dialogue act classification in domain-independent conversations using a deep recurrent neural network. In: COLING 2016, 26th International Conference on Computational Linguistics, pp. 2012–2021. Osaka (2016)
13. Kumar, V., Sridhar, R., Narayanan, S., Bangalore, S.: Enriching spoken language translation with dialog acts. In: Proceedings 46th Annual Meeting of the Association for Computational Linguistics Human Language Technology Short Paper - HLT 2008. p. 225, June 2008. https://doi.org/10.3115/1557690.1557755

14. Lee, J.Y., Dernoncourt, F.: Sequential short-text classification with recurrent and convolutional neural networks. In: NAACL 2016 (2016)

15. Li, D.: The pragmatic construction of word meaning in utterances. J. Chinese Lang. Comput. **18**(3), 121–137 (2012)

16. Liu, Y., Han, K., Tan, Z., Lei, Y.: Using context information for dialog act classification in DNN framework. In: Proceedings of the 2017 Conference on Empirical Methods in Natural Language Processing, pp. 2160–2168. ACL, Copenhagen, Denmark (2017)

17. Louwerse, M., Crossley, S.: Dialog act classification using N-gram algorithms. In: FLAIRS Conference on 2006. pp. 758–763. Melbourne Beach (2006)

18. Mandelbaum, A., Shalev, A.: Word embeddings and their use in sentence classification tasks. arXiv (2016)

19. Mikolov, T., Chen, K., Corrado, G., Dean, J.: Distributed representations of words and phrases and their compositionality. In: NIPS 2013 Proceedings of the 26th International Conference on Neural Information processing systems, pp. 3111–3119. Lake Tahoe (2013). https://doi.org/10.1162/jmlr.2003.3.4-5.951

20. Mikolov, T., Corrado, G., Chen, K., Dean, J.: Efficient estimation of word representations in vector space. arXiv (2013). https://doi.org/10.1162/153244303 322533223

21. Mikolov, T., Karafiát, M., Burget, L., Khudanpur, S.: Recurrent neural network based language model. In: INTERSPEECH, pp. 1045–1048. Makuhari (2010)

22. Papalampidi, P., Iosif, E., Potamianos, A.: Dialogue act semantic representation and classification using recurrent neural networks. In: SEMDIAL 2017 Workshop on the Semantics and Pragmatics of Dialogue, pp. 77–86, August 2017. https://doi.org/10.21437/SemDial.2017-9

23. Pennington, J., Socher, R., Manning, C.: Glove: global vectors for word representation. In: Proceedings of the 2014 Conference on Empirical Methods in Natural Language Processing, pp. 1532–1543 (2014). https://doi.org/10.3115/v1/D14-1162

24. Schegloff, E.A.: Sequence Oranization in Interaction: A Primer in Conversation Analysis I. Cambridge University Press, Cambridge (2007). https://doi.org/10.1017/CBO9780511791208

25. Searle, J.: Speech Acts: An Essay in the Philosophy of Language. Cambridge University Press, London (1969)

26. Stolcke, A., Ries, K., Coccaro, N., Shriberg, E., Bates, R., Jurafsky, D., Taylor, P., Martin, R., Van Ess-Dykema, C., Meteer, M.: Dialogue act modeling for automatic tagging and recognition of conversational speech. Comput. Linguist. **26**(3), 339–373 (2000). https://doi.org/10.1162/089120100561737

27. Surendran, D., Levow, G.A.: Dialog act tagging with support vector machines and hidden markov models. In: Interspeech 2006 9th International Conference on Spoken Language Processing, pp. 1950–1953. Pittsburgh (2006)

28. Webb, N., Hepple, M.: Dialogue act classification based on intra-utterance features. In: Proceedings of the AAAI Workshop on Spoken Language Understanding (2005)

29. Webb, N., Liu, T.: Investigating the portability of corpus-derived cue phrases for dialogue act classification. In: Proceedings of the 22nd International Conference on Computational Linguistics, vol. 1. pp. 977–984, Manchester, August 2008

Acceleration of Convolutional Networks Using Nanoscale Memristive Devices

Shruti R. Kulkarni, Anakha V. Babu, and Bipin Rajendran[✉]

Department of Electrical and Computer Engineering, NJIT, Newark, NJ, USA
{srk68,av442,bipin}@njit.edu

Abstract. We discuss a convolutional neural network for handwritten digit classification and its hardware acceleration as an inference engine using nanoscale memristive devices in the spike domain. We study the impact of device programming variability on the spiking neural network's (SNN) inference accuracy and benchmark its performance with an equivalent artificial neural network (ANN). We demonstrate optimization strategies to implement these networks with memristive devices with an on-off ratio as low as 10 and only 32 levels of resolution. Further, close to baseline accuracies can be maintained for the networks even if such memristive devices are used to duplicate the pre-determined kernel weights to enable parallel execution of the convolution operation.

Keywords: Spiking neural networks · Artificial neural networks
Non-volatile memory devices · Programming variability · Memristors

1 Introduction

Machine learning algorithms with impressive performance have been developed in the past five years for high-level cognitive tasks. Among these, convolution neural networks inspired by the hierarchical architecture of the visual cortex in the brain [1] have become state-of-the-art for various benchmark tasks [2–4]. However, in spite of several decades of research, the principles of information processing in the brain is not yet fully understood, and efforts to mimic its power efficiency and fault-tolerance in computing systems remain unfulfilled.

Spiking neural networks (SNNs) inspired by the dynamics of biological neurons [5] have also been proposed for numerous learning and actuation tasks. Signal communication in these networks occur below 1 kHz, similar to that in the human brain and may be crucial to achieve high energy efficiency [6,7]. Mimicking the event based computation and non-von Neumann architecture of the brain, several energy efficient CMOS chips such as the IBM's TrueNorth, INI Zurich's ROLLS, and Intel's Loihi have been demonstrated [8–10]. There are also several efforts to use emerging non-volatile memory (NVM) devices in neuromorphic hardwares for energy and area efficient designs [11,12].

Devices such as phase change memories (PCMs), spin-transfer-torque (STT) RAMs and Resistive RAMs (RRAMs) can be programmed to mimic conductivity modulation of biological synapses [12–16]. Numerical estimates of NVM

© Springer Nature Switzerland AG 2018
E. Pimenidis and C. Jayne (Eds.): EANN 2018, CCIS 893, pp. 240–251, 2018.
https://doi.org/10.1007/978-3-319-98204-5_20

based implementations suggest more than $25\times$ improvement in speed-up and up to $3000\times$ improvement in power compared to GPU based implementations [17]. NVM devices are particularly suited for processing-in-memory architectures, for instance the PipeLayer accelerator based on RRAMs for Convolution Neural Networks (CNNs) for training and testing showed significant speedup and energy efficiency over GPUs [18]. Other memristor based cross-bar implementations for CNNs showed that there is negligible loss in performance if the device has at least 4-bit resolution when used as inference engines [19]. Similarly, an extremely parallel architecture for accelerating deep CNNs has been proposed by leveraging the small size and high density of memristive devices [20]. Therefore, memristive devices offer one possible way to emulate brain's connectivity in hardware, if other non-ideal limitations of the devices can be mitigated. Good network performance can be obtained if the synaptic devices have linear and symmetric conductance response; but in reality, the device conductance is highly sensitive to programming variability and typically has finite on-off ratio with limited resolution [21]. It has been shown that conductance variability is a critical parameter that has to be optimized to maintain the ideal performance achievable in software simulations [22]. While the above mentioned efforts have studied the applicability of NVM devices with limited resolution for neural network realization, our work explores the impact of the programming variability of these devices on algorithmic performance.

We use a convolutional SNN for handwritten digit classification and describe a methodology of accelerating such a network on hardware using nanoscale memristive devices. We study the accuracy of this network as an inference engine when its synapses are implemented using memristive devices. We also benchmark the performance with respect to an equivalent second generation artificial neural network (ANN), where the trained SNN's accuracy is close to that of the trained ANN. Our results show that even at a very high device conductance variability, the accuracy for both the networks is within 1% of their respective baselines. We also study the realization of the convolution layer in our network as a complete parallel implementation by having a memristive device for each synaptic connection as opposed to having a shared synaptic array, as is done in the software and show that the parallel architecture helps in mitigating the effects of conductance variability on network's accuracy especially at higher variabilities. Thus, this work shows the high potential for realizing efficient inference engines based on event-triggered spiking networks implemented using memristive devices.

This paper is organized as follows. We first present the basic models of spiking neuron, synapses, the network architecture and the learning algorithms used in our work. We then introduce the optimization strategies to transform the weights obtained from software training into device compatible ranges and levels, including weight clipping and quantization. Next, we discuss two implementation schemes for hardware realization using memristive arrays for representing convolutional kernels, enabling the implementation of the convolution in a sequential and in a parallel manner. We compare the network performance as a function of the device programming variability in the results section. Finally, conclusions and future directions for this research are presented.

2   Spiking Neural Networks

SNNs are the artificial counterparts of the neurons found in biology that are inspired by the event based processing and signaling mechanisms of the brain [5]. While there have been numerous mathematical models explaining the various levels of details observed in biological neurons, the computationally simple leaky integrate and fire neuron (LIF) model typically suffices for capturing the essential dynamics of SNNs [23]. This model describes the neuron's membrane potential as a leaky integration of the incoming currents. The neuron issues a spike to its outgoing synapses when this potential exceeds a certain threshold.

The spikes issued by upstream neurons, propagate to the downstream neuron via synapses. The spikes are transformed into equivalent currents through the synaptic kernels, which are modeled using a double decaying exponential kernel.

3   Network Architecture

A three layered convolutional network to classify handwritten digits from the MNIST database as proposed in [24] is used in this study (Fig. 1). The input layer has 784 LIF neurons, each taking input from a pixel of the MNIST image, which is of size 28×28. This layer transforms the pixels into a stream of spikes, with frequency sub-linearly proportional to the pixel amplitude, for a duration T. The input layer spikes are propagated to a convolution layer which has 12 feature maps. Each of these 12 maps is associated with a 3×3 convolution kernel, whose weights are set *a priori*. The fixed weight kernels were designed to identify the edges and corners from the input image into different channels as a pre-processing step, similar to the approach in [25]. The output layer has 10 neurons, each corresponding to a particular digit from 0 to 9. All the neurons from the hidden convolution layer connect in an all-to-all manner to the output layer neurons.

The spike based supervised learning rule NormAD was used to train the synaptic weights of the feed-forward synapses of the output layer neurons of the SNN [26]. The output layer of the SNN also has fixed strength lateral inhibitory connections that prevent the non-label neurons from spiking for a given input. For making a decision on the label of the input image, we used a metric that calculates the correlation between the precise timing information of a spike train with respect to a reference spike train, as described in [27]. The SNN was trained in software with the spike based NormAD rule for 20 training epochs with the 60,000 images from the MNIST training set and it achieved a test set accuracy of 98.17%. Each image is presented for a duration $T = 100$ ms. Further details on the fine tuning of the learning rule parameters and network optimization can be found in [28].

The reference ANN consists of ReLU units in the hidden layer and softmax units in the output layer. It is trained with the standard backpropagation rule for 20 epochs. The use of softmax neurons in the output layer along with cross-entropy cost function helped in improving the network's classification accuracy over using ReLU neurons throughout the network. The neuron with the highest activation output is declared the winner in the ANN. The MNIST test set accuracy achieved by the ANN is 98.10%.

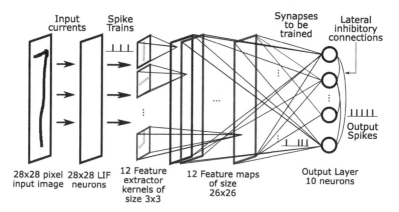

Fig. 1. The SNN architecture used for handwritten digit classification: the spike trains from the input layer with 28×28 neurons are spatially convolved with twelve filters of size 3×3, resulting in the twelve feature maps (size 26×26). The synapses connecting the 8112 convolution layer neurons and the 10 output layer neurons are tuned during training. There is a fixed winner-take-all (WTA) lateral inhibition between the neurons in the output layer. The architecture of the reference ANN is similar to the above SNN, with the neurons employing ReLU and softmax activation functions; the ANN output layer does not have any lateral inhibition.

4   Network Optimization for Hardware

We now discuss the network optimization strategies to translate the software design for energy and area efficient neuromorphic hardware platforms. Typically, such platforms have limited precision for weight storage. Previous efforts to study the impact of low-precision weights in a digital realization have shown close to 5% drop in accuracy with respect to the baseline even with 5-bits of precision for synaptic weights [29].

Memristive devices can also be used as synaptic weights in crossbar neuromorphic platforms [30]. Even though the device conductance is an analog value, the granularity to which a device can be programmed to a particular level is typically limited, resulting in a finite number of levels within the dynamic range of the device. In the following subsections we discuss our study on the impact of realizing synaptic weights with memristive devices, with an on-off ratio of 10 and a resolution of approximately 32 levels (or 5-bits). These characteristics are typical of several experimental memristive devices today.

4.1 Restricted ON-OFF Ratios of Synaptic Weights

To study the impact of using memristive devices as synapses in our neural network, we measure the inference accuracy by limiting the range of weight values that were obtained after training the networks in software. The learned weight values of the neural networks trained in software were limited to have an on-off ratio of 10. Figure 2 shows histogram of the software trained weights and the histogram after clipping them to have an on-off ratio of 10. Table 1 shows the accuracy and the number of non-zero connections in the feed-forward fully-connected layer of the SNN and ANN, before and after removing the insignificant weight values. It can be seen that even after eliminating more than 50% of the trained weights from the two networks, the drop in test accuracy is negligible, indicating that sparsely connected networks are capable of delivering close to baseline accuracies.

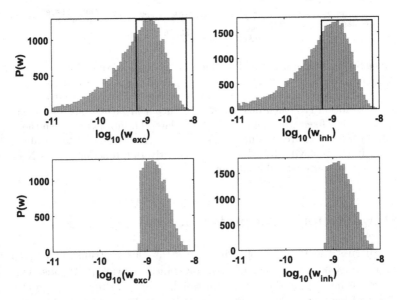

Fig. 2. Software trained weights (upper panel) having a large range of values, are clipped such that the ratio of maximum value to minimum value (on-off ratio) of the excitatory and inhibitory weights is restricted to 10. This range of values resulted in the test accuracy to drop to 98.07% from the baseline value of 98.17% in the SNN. For the ANN, there was no drop in the test accuracy.

4.2 Hardware Architecture

While the synapses between the fully-connected layers can be implemented using a cross-bar array in a straight-forward manner [17], the implementation of the convolution operation is the focus of our attention in this study. A convolution layer in a neural network uses a weight matrix (also called convolution kernels),

Table 1. Test accuracy of the SNN and the ANN when the weights are clipped to an on/off ratio of 10 after training.

Network	SNN	ANN
Non-zero synapses in baseline network	75,000	78,000
Baseline accuracy on MNIST test set	98.17%	98.10%
Non-zero synapses in the new network	30,000	27,000
Inference accuracy in the new network	98.07%	98.10%

whose element values are tuned to extract particular features from the input image. The software implementation of the convolution sequentially repeats the matrix across the entire input, and a dot-product of the input overlap with the weight matrix is computed. Although this computation can be accelerated on a GPU (graphical processing unit), by scheduling the computation of convolution for elements and kernels in a concurrent manner, the limited number of computation cores always limits the degree of parallelization possible, thereby limiting the overall speed.

There have been some recent efforts to implement the convolution operation using NVM based neuromorphic hardware [18–20,31,32]. Here, we study the impact of using the analog conductance levels in memristors, with two devices in a differential configuration per synaptic weight to realize the excitatory and inhibitory weight connections [19,30].

4.3   Synapses Using Memristive Devices

In order to represent positive and negative weights (w), we use two memristive devices with conductances G^+ and G^- per synapse [30], such that

$$w = k(G^+ - G^-) \tag{1}$$

Appropriate scaling factor (k) is used to translate the device conductances to the range of software trained weights. Most memristive devices also exhibit gradual conductance change in one direction; hence we assume a unidirectional programming scheme [33]. The device is assumed to have an on-off ratio of 10 and roughly 32 levels of programming resolution for realizing the synaptic weights. In this scheme, either G^+ (G^-) will be programmed to any of the allowed 32 conductance states, depending on the sign of the software weights at every synapse. For the fully connected layer in SNN and ANN, including the inhibitory weights of SNN, when the synaptic weights are zeros, both G^+ and G^- are programmed to the minimum conductance in the linear regime.

4.4   Sequential and Parallel Convolution

We now discuss two hardware architectural schemes for implementing the convolution layer using memristive cross-bar arrays and illustrate the architectures

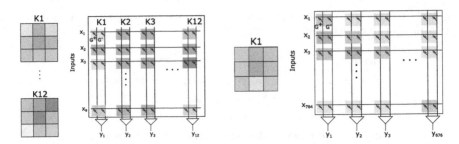

Fig. 3. (Left) Sequential convolution in memristive crossbar array with $9 \times 12 \times 2$ devices to represent the 12 kernels used in the convolution layer, each having a 3×3 sized weight matrix. These matrices are unrolled as vectors of size (9×1). The inputs need to be presented in sections of 9 elements to obtain the output of the convolution operation. (Right) Parallel convolution in memristive crossbar array with $784 \times 676 \times 2$ devices. Each neuron in the convolution layer has 9 incoming synapses, so every column in the array has only 9 active connections. The cross-points in gray are inactive connections.

using a convolution operation performed between an input image of size 28×28 with a 3×3 kernel, resulting in a output matrix with 26×26 elements. In the first scheme, only the 9 unique values in the convolution kernel are represented using 9×2 memristive devices [19,31]. Figure 3 (left) shows the architecture of crossbar array, where the 12 kernels are laid out in 12 columns, each of length 9. Here, the inputs x_i need to be presented to the array in batches of 9 elements at a time. The outputs y_j, for each kernel are obtained sequentially, by the application of Kirchhoff's law. A voltage signal proportional to the input quantities x_i are applied on the horizontal wires, and the resulting current through the synapses are accumulated by the sense amplifiers on the vertical wires. While this scheme uses only $9 \times 12 \times 2 = 216$ devices for the array, 676 *sequential* cycles are needed to complete the convolution across all kernels for each MNIST image.

In the second scheme, a larger memristive array is used so that all the convolution operations are completed in parallel. In this array, every connection between the input and output has a synaptic device associated with it [20]. Figure 3 (right) shows the architecture of the crossbar array with 784×676 synapses for a single convolution kernel output map. Such arrays need to be repeated for multiple kernel output maps. This scheme requires $2 \times 784 \times 676 = 1,059,968$ memristive synaptic devices per convolution output map. However, since each neuron in the convolution layer connects to only 9 inputs, the active number of connections in the network are only $9 \times 676 = 6084$, resulting in a sparsity of $\sim 1\%$. With this scheme, we can achieve a speedup of $676\times$ over the sequential realization. The implications of using multiple memristors, which are known to have conductance variabilities, for every synaptic connection as opposed to having a shared array is discussed in the Sect. 5.

4.5   Programming Variability

Memristive devices are non-ideal, exhibiting significant programming variability, which may affect the network performance [21,22]. The impact of programming variability in the network performance is studied using the parameter σ/B, where σ is the standard deviation and B is the bin-width of the conductance states obtained during programming. In order to emulate the programming variability of the devices in the simulation, a zero mean Gaussian noise of standard deviation σ is added to the programmed device conductance. We study the network's inference accuracy for different programming variability with $\sigma/B = 0.5$, $0.8, 1$ and 1.5. The programming variability used in this study is comparable with experimental reported values for typical memristive devices [34]. As the programming variability parameter is increased from $\sigma/B = 0.5$ to 1.5, the conductance spills over to the neighboring bins, thereby potentially affecting the network's inference accuracy.

In our simulations, this conductance variability is included in the synapses for all the layers in the parallel and sequential convolution networks. Further, we assume that the inactive devices in the convolution layers can be programmed to $0.1 \times G_{min}$ so as to minimize the effect of programming variability in the convolution kernel. Also, the programming variability associated with these inactive devices is assumed to be one-tenth of σ/B. These assumptions are based on the experimentally observed characteristics of emerging memories, whose off-state conductance can be at least 10 times lower than the typical analog range used for neuromorphic weight storage [34,35].

5   Results

Our SNN and ANN are implemented in CUDA-C and C, respectively. The software baseline response of SNN and ANN are at 98.07% and 98.10% inference accuracy, respectively when the weights are clipped. On translating these clipped weights to the allowed conductance states of the memristive device without any variability, the inference response drops slightly to 97.99% for SNN while it remains the same for ANN. These accuracies are used as the device baseline for analyzing the response for different programming variability.

5.1   Sequential Convolution

The response of sequential convolution architecture of SNN and ANN after introducing programming variability to the memristive devices is shown in Fig. 4 (left). It can be seen that as the variability is increased, ANN suffers slightly less degradation in accuracy when compared to SNN. The input currents corresponding to those images that were misclassified by SNN ($\sigma/B = 1.5$) but correctly classified by ANN ($\sigma/B = 1.5$) and the device baseline networks of SNN ($\sigma/B = 0$) and ANN ($\sigma/B = 0$) is analyzed in Fig. 5. The neuronal input current in the SNN is obtained as an integrated value over a duration of $T = 100\,\mathrm{ms}$,

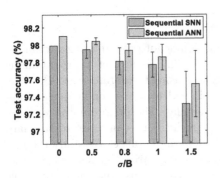 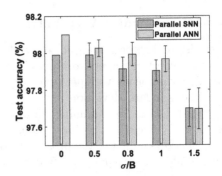

Fig. 4. Accuracy of the spiking and non-spiking networks for sequential (left) and parallel (right) convolution as a function of the device conductance level variations, defined as the ratio σ/B, where σ is the standard deviation of the zero mean Gaussian noise and B is the bin-width of the conductance levels. In both the cases, the average classification accuracy of the SNN is close to that of the ANN within 0.1%.

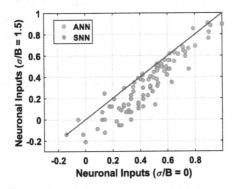

Fig. 5. The incoming currents to the output layer neurons of both SNN and ANN for $\sigma/B = 1.5$. The x-axis corresponds to the baseline network without any programming variability ($\sigma/B = 0$), while the y-axis represents the networks with variability $\sigma/B = 1.5$. It can be seen that for SNN, input currents deviated more from the baseline when compared to ANN, resulting in the slightly higher accuracy drop.

while for the ANN it is the instantaneous DC value of the weighted inputs to each neurons. It can be seen that the deviation in the ANN from the baseline network having no variability is lesser compared to that of the SNN, which explains the lesser degradation in the accuracy of ANN (Fig. 4).

5.2   Parallel Convolution

High density packing of the memristive devices can be leveraged to implement parallel convolution giving significant speed up in hardware [20]. The convolution operation for the SNNs can also be implemented in parallel using the network architecture in Fig. 3 (right). Similar to the sequential convolution implementation, even in the parallel network, the SNN's accuracy was close to that of the ANN within ~0.1% (see Fig. 4 (right)). As can be seen from Fig. 6, the inference accuracy of the two schemes (parallel and sequential) is comparable, although the parallel implementation shows slightly better accuracies at high device programming variability. The slightly better performance with the parallel convolution architecture may be due to the averaging and compensating effect of memristive devices in determining the output of the convolutional operation. In order to obtain this performance, it is important that the inactive devices in the array are programmed to low conductance values, below the normal dynamic range used for synaptic weight representation.

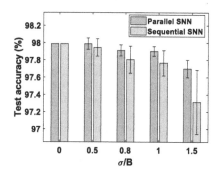 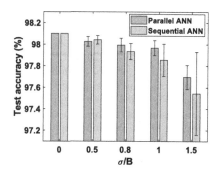

Fig. 6. Comparison of networks' inference accuracy for sequential and parallel convolution architectures with memristive arrays in SNNs (left) and ANNs (right).

6   Conclusion

We presented a convolutional neural network in both spiking (SNN) and non-spiking (ANN) versions, realized using memristive cross-bar arrays. The networks are trained in software in the full-precision mode and their synaptic weights are then optimized for designing an inference engine using memristive devices. Our simulations suggest that optimization strategies such as weight clipping and realization of the convolution operation in parallel using memristive arrays can result in implementations with close-to-baseline accuracies. As our SNN performs nearly as well as the reference ANN on memristive hardware with conductance variability, it shows potential for realizing energy efficient neuromorphic platforms using SNNs. The optimization of the architectural schemes to implement on-chip learning using memristive networks is identified as a main topic for future work.

Acknowledgments. This research was supported in part by the CAMPUSENSE project grant from CISCO Systems Inc, the Semiconductor Research Corporation (2016-SD-2717), and the National Science Foundation grant 1710009.

References

1. Hubel, D., Wiesel, T.: Receptive fields and functional architecture of monkey striate cortex. J. Physiol. **195**(1), 215–243 (1968)
2. Lecun, Y., et al.: Gradient-based learning applied to document recognition. Proc. IEEE **86**(11), 2278–2324 (1998)
3. Krizhevsky, A., Sutskever, I., Hinton, G.: Imagenet classification with deep convolutional neural networks. In: Advances in neural information processing systems, pp. 1097–1105 (2012)
4. Szegedy, C., et al.: Inception-v4, Inception-ResNet and the impact of residual connections on learning. In: AAAI, vol. 4, p. 12 (2017)
5. Maass, W.: Networks of spiking neurons: the third generation of neural network models. Neural Netw. **10**(9), 1659–1671 (1997)
6. Wang, B., et al.: Firing frequency maxima of fast-spiking neurons in human, monkey, and mouse neocortex. Front. Cell. Neurosci. **10**, 239 (2016). 27803650 [pmid]
7. Han, B., Sengupta, A., Roy, K.: On the energy benefits of spiking deep neural networks: a case study. In: 2016 International Joint Conference on Neural Networks (IJCNN), pp. 971–976. IEEE (2016)
8. Merolla, P.A., et al.: A million spiking-neuron integrated circuit with a scalable communication network and interface. Science **345**(6197), 668–673 (2014)
9. Qiao, N., et al.: A reconfigurable on-line learning spiking neuromorphic processor comprising 256 neurons and 128k synapses. Front. Neurosci. **9**, 141 (2015)
10. Davies, M., et al.: Loihi: a neuromorphic manycore processor with on-chip learning. IEEE Micro **38**(1), 82–99 (2018)
11. Kim, S., et al.: NVM neuromorphic core with 64k-cell (256-by-256) phase change memory synaptic array with on-chip neuron circuits for continuous In-Situ learning. In: 2015 IEEE International Electron Devices Meeting (IEDM), pp. 17.1.1–17.1.4, December 2015
12. Burr, G.W., et al.: Neuromorphic computing using non-volatile memory. Adv. Phys. X **2**(1), 89–124 (2017)
13. Rajendran, B., Alibart, F.: Neuromorphic computing based on emerging memory technologies. IEEE J. Emerg. Sel. Top. Circ. Syst. **6**(2), 198–211 (2016)
14. Kuzum, D., Yu, S., Wong, P.: Synaptic electronics: materials, devices and applications. Nanotechnology **24**(38), 382001 (2013)
15. Jo, S.H., et al.: Nanoscale memristor device as synapse in neuromorphic systems. Nano Lett. **10**(4), 1297–1301 (2010). PMID: 20192230
16. Jackson, B.L., et al.: Nanoscale electronic synapses using phase change devices. J. Emerg. Technol. Comput. Syst. **9**(2), 12 (2013)
17. Burr, G.W., et al.: Large-scale neural networks implemented with non-volatile memory as the synaptic weight element: comparative performance analysis (accuracy, speed, and power). In: 2015 IEEE International Electron Devices Meeting (IEDM), pp. 4.4.1–4.4.4, December 2015
18. Song, L., et al.: PipeLayer: a pipelined ReRAM-based accelerator for deep learning. In: 2017 IEEE International Symposium on High Performance Computer Architecture (HPCA), pp. 541–552, February 2017
19. Yakopcic, C., Alom, Z., Taha, T.: Memristor crossbar deep network implementation based on a convolutional neural network. In: International Joint Conference on Neural Networks (2016)
20. Yakopcic, C., Alom, Z., Taha, T.: Extremely parallel memristor crossbar architecture for convolutional neural network implementation. In: 2017 International Joint Conference on Neural Networks (IJCNN), pp. 1696–1703. IEEE (2017)

21. Chen, P.Y., et al.: Mitigating effects of non-ideal synaptic device characteristics for on-chip learning. In: 2015 IEEE/ACM International Conference on Computer-Aided Design (ICCAD), November 2015
22. Babu, A.V., Rajendran, B.: Stochastic deep learning in memristive networks. In: 2017 24th IEEE International Conference on Electronics, Circuits and Systems (ICECS), pp. 214–217, December 2017
23. Abbott, L.: Lapicque's introduction of the integrate-and-fire model neuron (1907). Brain Res. Bull. **50**, 303–304 (1999)
24. Kulkarni, S.R., Alexiades, J.M., Rajendran, B.: Learning and real-time classification of hand-written digits with spiking neural networks. In: 2017 24th IEEE International Conference on Electronics, Circuits and Systems (ICECS), pp. 128–131, December 2017
25. Calderón, A., Roa, S., Victorino, J.: Handwritten digit recognition using convolutional neural networks and Gabor filters. In: Proceedings of International Congress on Computational Intelligence (2003)
26. Anwani, N., Rajendran, B.: NormAD - normalized approximate descent based supervised learning rule for spiking neurons. In: International Joint Conference on Neural Networks, pp. 1–8, July 2015
27. Schreiber, S., et al.: A new correlation-based measure of spike timing reliability. Neurocomputing **52**, 925–931 (2003)
28. Kulkarni, S.R., Rajendran, B.: Spiking neural networks for handwritten digit recognition-supervised learning and network optimization. Neural Netw. **103**, 118–127 (2018)
29. Stromatias, E., et al.: Robustness of spiking deep belief networks to noise and reduced bit precision of neuro-inspired hardware platforms. Front. Neurosci. **9**, 222 (2015)
30. Suri, M., et al.: Phase change memory as synapse for ultra-dense neuromorphic systems: application to complex visual pattern extraction. In: 2011 International Electron Devices Meeting, pp. 4.4.1–4.4.4, December 2011
31. Gokmen, T., Onen, M., Haensch, W.: Training deep convolutional neural networks with resistive cross-point devices. arXiv preprint arXiv:1705.08014 (2017)
32. Garbin, D., et al.: HfO2-based OxRAM devices as synapses for convolutional neural networks. IEEE Trans. Electron Devices **62**(8), 2494–2501 (2015)
33. Lim, S., et al.: Adaptive learning rule for hardware-based deep neural networks using electronic synapse devices. ArXiv e-prints arXiv:1707.06381v2, July 2017
34. Boybat, I., et al.: Neuromorphic computing with multi-memristive synapses. ArXiv e-prints, November 2017
35. Panwar, N., Rajendran, B., Ganguly, U.: Arbitrary spike time dependent plasticity (STDP) in memristor by analog waveform engineering. IEEE Electron Device Lett. **38**(6), 740–743 (2017)

Comparison of Asymmetric and Symmetric Neural Networks with Gabor Filters

Naohiro Ishii[1]([⊠]), Toshinori Deguchi[2], Masashi Kawaguchi[3], and Hiroshi Sasaki[4]

[1] Aichi Institute of Technology, Toyota, Japan
ishii@aitech.ac.jp
[2] Gifu National College of Technology, Gifu, Japan
deguchi@gifu-nct.ac.jp
[3] Suzuka National College of Technology, Mie, Japan
masashi@elec.suzukact.ac.jp
[4] Fukui University of Technology, Fukui, Japan
hsasaki@fukui-ut.ac.jp

Abstract. Visual motion information is useful for many complex tasks in the biological and robotic systems. Models for motion processing in the biological systems have been studied to use conventional symmetric quadrature functions with Gabor filters. This paper proposes a model of the another bio-inspired asymmetric neural networks. The prominent features are the nonlinear characteristics as the squaring and rectification functions, which are observed in the retinal and visual cortex networks. In this paper, the asymmetric network with Gabor filters is compared with that of the conventional symmetric networks. It is shown that the biological asymmetric network with nonlinearities is effective for detecting the inputted phase information and directional movements from the network computations. The responses to the frequency characteristics and to the complex motion stimulus are computed in the asymmetric networks, which are not derived for the conventional energy model.

Keywords: Asymmetric neural network · Gabor filter
Motion analysis in frequency domain · Linear and nonlinear pathways

1 Introduction

Motion information plays an important role for the processing complex tasks in the biological and robotic systems. To estimate the visual motion, sensory biological information models have been studied [1–4]. Reichard [1] evaluated the sensory information by the auto-correlations in the neural networks. Physiological studies show the directional sensitive functions in the brain cortex [3–6]. From these studies, the image information is characterized by orientation functions in space-time, which are called Gabor filters. By using Gabor filters, a symmetric network model was developed for the motion detection, which is called energy model [2]. In the biological visual systems, the nonlinear characteristics are observed as the squaring function and rectification function, which exist in the retina and the visual cortex networks [2, 3, 5–7]. Though the visual

© Springer Nature Switzerland AG 2018
E. Pimenidis and C. Jayne (Eds.): EANN 2018, CCIS 893, pp. 252–263, 2018.
https://doi.org/10.1007/978-3-319-98204-5_21

cortex shows an extended asymmetric networks, the basic network is based on the proposed asymmetric networks here. Thus, in this paper, an another bio-inspired model is proposed for the motion detection by using these nonlinear characteristics. These nonlinear functions are taken in the asymmetrical neural networks [12, 14, 15]. The proposed asymmetric network is compared with the conventional energy model under the same simple inputted stimulus conditions. The phase information has been studied to be useful for the visual complex tasks. It is shown that the asymmetric network with nonlinearities is effective for the detection of the phase information in the frequency domain, which is not extracted using conventional energy model. Finally, complex stimulus are applied to the asymmetric network with Gabor filters.

2 Symmetric Networks with Gabor Filters

A symmetric network with Gabor filter were proposed by Adelson and Bergen [2] as the energy model of the perception of the visual motion. Their model is extensively applied to the motion detection and physiological models as the fundamental model of the neural networks. For the application of motion detection based on the energy model, Heeger [7, 8] presented an elegant model that computes velocities through the spatio-temporal integration of the outputs of Gabor motion energy filters. Grzywacz and Yuile [3] presented an integrated model, qualitatively consistent with physiology, that postulates two stages for cortical velocity estimate: the first measuring motion energies from the moving stimulus and the second estimating velocity from these energies.

Adelson and Bergen [2] suggested that the following filter called Gabor filter can detect the image motion by the orientation of the space–time. The Gabor filter is described as [3]

$$
f(x, t : \Omega, n, \Omega_t, \sigma, \sigma_t) = \frac{1}{(2\pi)^{3/2}\sigma^2(\sigma_t)} \exp\left(-\frac{|x|^2}{2\sigma^2}\right) \exp\left(-\frac{t^2}{2\sigma_t^2}\right)
$$
$$
\times \exp(-i\Omega n \cdot x)\exp(-i\Omega_t t) \tag{1}
$$

where x and t are a spatial location in the image and time, respectively, $\sigma > 0$, $\sigma_t > 0$, Ω and Ω_t are scalar parameters, and $n = (\cos\theta, \sin\theta)$ is a unit vector.

This filter is used also by Heeger [7] as the real part for the cosine-phase Gabor filter

$$
Gab_{\cos}(x, t) = \frac{1}{(2\pi)^{3/2}\sigma^2(\sigma_t)} \exp\left(-\frac{|x|^2}{2\sigma^2}\right) \exp\left(-\frac{t^2}{2\sigma_t^2}\right)
$$
$$
\times \cos(\Omega n \cdot x + \Omega_t t) \tag{2}
$$

and its imaginary part is minus the sine-phase Gabor filter

$$Gab_{\sin}(x,t) = \frac{1}{(2\pi)^{3/2}\sigma^2(\sigma_t)}\exp\left(-\frac{|x|^2}{2\sigma^2}\right)\exp\left(-\frac{t^2}{2\sigma_t^2}\right)$$
$$\times \sin(\Omega n \cdot x + \Omega_t t)$$

(3)

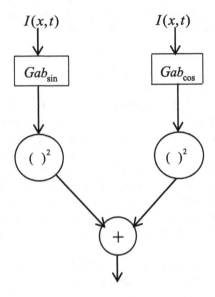

Fig. 1. Energy model with Gabor filter

The energy model with Gabor filters ($Gab_{\sin}$ and $Gab_{\cos}$) is shown in Fig. 1.

The responses of the directionally selective cells in the primary visual cortex to an image is modeled as

$$Res = |f(x,t : \Omega, n, \Omega_t, \sigma, \sigma_t) * I(x,t)|^2$$

(4)

where * denotes convolution and $I(x,t)$ shows a two dimensional image in the spatial and time domain. The equation is called motion energy [2].

The Fourier transform of the Eq. (1) is computed [3] as

$$F(\omega, \omega_t : \Omega, n, \Omega_t, \sigma, \sigma_t) = \frac{1}{(2\pi)^{3/2}\sigma^2(\sigma_t)}\exp\left(-\frac{(n \cdot \omega - \Omega)^2\sigma^2}{2}\right)$$
$$\times \exp\left(-\frac{(\omega_t - \Omega_t)^2\sigma_t^2}{2}\right)$$

(5)

where x and t are a spatial location in the image and time, respectively. Notations Ω and Ω_t are scalar parameters in the frequency domain. These notations are often used as ω and ω_t, respectively. To compare the output response of the energy model of the symmetric network with that of the asymmetric network, the input stimulus is assumed here. The simple case of the input is considered here as setting $n = 1 (\theta = 0)$ in the Eqs. (1), (2), (4) and (5). Further, a simple motion stimulus is inputted to the conventional energy model with Gabor filter, which is the symmetric network with Gabor filter. The moving sinusoidal stimulus input is adopted under the conditions of the stimulus to be $n = 1 (\theta = 0)$ [3] as follows,

$$
\begin{aligned}
I(x, t) &= I_1 \sin(\lambda(x - vt)) \\
&= \frac{I_1}{2j} (e^{+j(\lambda(x-vt))} - e^{-j(\lambda(x-vt))})
\end{aligned}
\tag{6}
$$

Fourier transform of the moving stimulus in the Eq. (6) becomes

$$
F(I(x,t)) = \iint \frac{I_1}{2j} (e^{+j(\lambda(x-vt))} - e^{-j(\lambda(x-vt))}) e^{-j(\omega x + \omega_t t)} dx dt
\tag{7}
$$

To derive the output response of the energy model of the symmetric network, the convolution of the symmetric network and the moving input stimulus is obtained in the frequency domain as follows in the Eq. (8),

$$
\begin{aligned}
&F(\omega, \omega_t : \Omega, n, \Omega_t, \sigma, \sigma_t) \cdot F(I(x,t)) \\
&= \frac{1}{(2\pi)^{3/2}} \cdot \frac{I_1}{2j} \left\{ exp(-\frac{(\lambda - \Omega)^2 \sigma^2}{2}) \exp(-\frac{(\lambda v + \Omega_t)\sigma_t^2}{2}) - \exp(-\frac{(\lambda + \Omega)^2 \sigma^2}{2}) \exp(-\frac{(\lambda v - \Omega_t)\sigma_t^2}{2}) \right\} \\
&= \frac{1}{(2\pi)^{3/2}} \frac{j}{2} (A - C)
\end{aligned}
\tag{8}
$$

where the following notations A and C are used.

$$
\begin{aligned}
A &= \exp(-\frac{(\lambda - \Omega)^2 \sigma^2}{2}) \cdot \exp(-\frac{(\lambda v + \Omega_t)\sigma_t^2}{2}) \text{ and} \\
C &= \exp(-\frac{(\lambda + \Omega)^2 \sigma^2}{2}) \cdot \exp(-\frac{(\lambda v - \Omega_t)\sigma_t^2}{2})
\end{aligned}
$$

Thus, the total response of the Eq. (4) becomes in the frequency domain.

$$
|F(\omega, \omega_t : \Omega, n, \Omega_t, \sigma, \sigma_t) \cdot F(I(x,t))|^2 = \frac{1}{(2\pi)^3} \frac{1}{4} (A - C)^2
\tag{9}
$$

3   Asymmetric Neural Networks

3.1   Background of Asymmetric Neural Networks

In the biological neural networks, the structure of the network, is closely related to the functions of the network. Naka et al. [12] presented a simplified, but essential networks of catfish inner retina as shown in Fig. 2. Visual perception is carried out firstly in the retinal neural network as the special processing between neurons. The following asymmetric neural network is extracted from the catfish retinal network [12]. The asymmetric structure network with a quadratic nonlinearity is shown in Fig. 1, which composes of the pathway from the bipolar cell B to the amacrine cell N and that from the bipolar cell B, via the amacrine cell C to the N [11, 12].

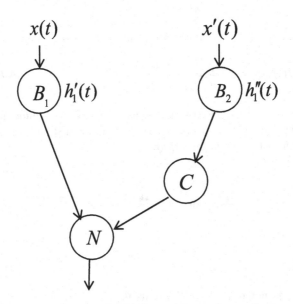

Fig. 2. Asymmetric network with linear and squaring nonlinear pathways

Figure 2 shows a network which plays an important role in the movement perception as the fundamental network. It is shown that N cell response is realized by a linear filter, which is composed of a differentiation filter followed by a low-pass filter. Thus, the asymmetric network in Fig. 2 is composed of a linear pathway and a nonlinear pathway with the cell C, which works as a squaring function.

3.2   Application of Asymmetric Networks to Biological Neural Networks

Here, we present an example of layered neural network in Fig. 3(a), which is developed from the neural network in the brain cortex [6]. Figure 3 is a connected network model of V1 followed by MT, where V1 is the front part of the total network, while MT is the rear part of it. Figure 3(a) is transformed to the approximated one as follows.

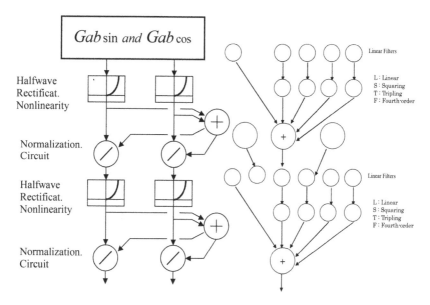

Fig. 3. (a) Neural model of V1 and MT [6], (b) Approximated model shown in (a)

The half-wave rectification in Fig. 3(a) is approximated in the following equation.

$$f(x) = \frac{1}{1 + e^{-\eta(x-\theta)}} \tag{10}$$

By Taylor expansion of the Eq. (10) at $x = \theta$, the Eq. (11) is derived as follows,

$$f(x)_{x=\theta} = f(\theta) + f'(\theta)(x - \theta) + \frac{1}{2!}f''(\theta)(x - \theta)^2 + \cdots$$
$$= \frac{1}{2} + \frac{\eta}{4}(x - \theta) + \frac{1}{2!}(-\frac{\eta^2}{4} + \frac{\eta^2 e^{-\eta\theta}}{2})(x - \theta)^2 + \cdots \tag{11}$$

Though the network developed by the approximated network with Tailor expansion in Fig. 3(b), shows an extended asymmetric nonlinear network, the basic network as the asymmetric network is shown in Fig. 4 in the following [15].

The left pathway with $Gab_{\sin}$ of the asymmetric network in Fig. 4 consists of the convolution of the $Gab_{\sin}$ and input stimulus $I(x, t)$. Thus, the Fourier transform of the convolution becomes the product of the Fourier transforms of $Gab_{\sin}$ and $I(x, t)$. Thus, the Fourier transform of $Gab_{\sin}$ of the Eq. (3) becomes

$$F(Gab_{\sin}) = \frac{1}{(2\pi)^{3/2}} \frac{j}{2} \left\{ \exp(-\frac{(\omega - \Omega)^2\sigma^2}{2}) \exp(-\frac{(\omega_t - \Omega_t)^2\sigma_t^2}{2}) + \exp(-\frac{(\omega + \Omega)^2\sigma^2}{2}) \exp(-\frac{(\omega_t + \Omega_t)^2\sigma_t^2}{2}) \right\} \tag{12}$$

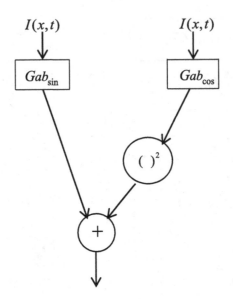

Fig. 4. Asymmetric neural network with Gabor filters

The response of the left pathway with Gab_{sin} becomes to be convolution in the frequency domain,

$$F(Gab_{sin}) * F(I(x, t)) \tag{13}$$

The Eq. (13) of the left pathway becomes the Eq. (14)

$$\frac{1}{(2\pi)^{3/2}} \frac{I_1 j}{2} \left\{ 2 \exp(-\frac{(\lambda - \Omega)^2 \sigma^2}{2}) \exp(-\frac{(\lambda v + \Omega_t)\sigma_t^2}{2}) - 2 \exp(-\frac{(\lambda + \Omega)^2 \sigma^2}{2}) \exp(-\frac{(\lambda v - \Omega_t)\sigma_t^2}{2}) \right\}$$
$$= \frac{I_1 j}{(2\pi)^{3/2}} (A - C) \tag{14}$$

The response of the right pathway with Gab_{cos} filter becomes

$$\frac{1}{(2\pi)^{3/2}} \frac{I_1}{2} \left\{ \exp(-\frac{(\lambda - \Omega)^2 \sigma^2}{2}) \exp(-\frac{(\lambda v + \Omega_t)\sigma_t^2}{2}) + \exp(-\frac{(\lambda + \Omega)^2 \sigma^2}{2}) \exp(-\frac{(\lambda v - \Omega_t)\sigma_t^2}{2}) \right.$$
$$\left. - \exp(-\frac{(\lambda + \Omega)\sigma^2}{2}) \exp(-\frac{(\lambda v - \Omega_t)\sigma_t^2}{2}) - \exp(\frac{(\lambda - \Omega)^2 \sigma^2}{2}) \exp(-\frac{(\lambda v + \Omega_t)\sigma_t^2}{2}) \right\} = 0 \tag{15}$$

Since the Eq. (15) of the right pathway with Gab_{cos} is zero, the response of the asymmetric network is shown to be the value of the left pathway with Gab_{sin} filter as shown in the Eq. (14). Since the response value of the energy model becomes

$\frac{1}{(2\pi)^3}\frac{1}{4}(A-C)^2$ in the Eq. (9), while that of the asymmetric network shows $(\frac{l_i j}{(2\pi)^{3/2}})(A-C)$ in the value. This shows $(A-C)$ is the common same factor to show the responses to the motion velocity stimulus for the symmetric and the asymmetric networks. In the next sections, different responses are shown between the symmetric and the asymmetric networks.

4 Extraction of Frequency Characteristics in the Asymmetric Networks

Prof. Heeger derived the output of the conventional energy model with Gabor filters [7], in the frequency domain which shows only the Gabor energy without the phase-information as shown in the following. To the given sine-wave stimulus with the angular frequency $\bar{\omega}$ and the phase ϕ, the squared-output of the sine-phase Gabor filter convolved with the sine-wave input is derived [7] as follows,

Input stimulus: $\sin(2\pi\bar{\omega}x + \phi)$

Output of the sine-phase Gabor filter in the Eq. (16):

$$\int_{-\infty}^{\infty} |G_s(\omega_0,\sigma) * \sin(2\pi\bar{\omega}x+\phi)|^2 dx = \int_{-\infty}^{\infty} |F\{G_s(\omega_0,\sigma)\}F\{\sin(2\pi\bar{\omega}x+\phi)\}|^2 d\omega$$
$$= (1/8)[exp[-2\pi^2\sigma^2(\bar{\omega}-\omega_0)^2 - exp[-2\pi^2\sigma^2(\bar{\omega}+\omega_0)^2]^2$$

(16)

while the output of the cosine-Gabor filter becomes the Eq. (17)

$$\int_{-\infty}^{\infty} |G_c(\omega_0,\sigma) * \cos(2\pi\bar{\omega}x+\phi)|^2 dx$$
$$= (1/8)[exp[-2\pi^2\sigma^2(\bar{\omega}-\omega_0)^2 + exp[-2\pi^2\sigma^2(\bar{\omega}+\omega_0)^2]^2$$

(17)

Combining Eqs. (16) and (17) becomes the output of the energy model in the frequency domain which gives the phase-independent energy as follows [7],

$$(1/4)\{exp[-4\pi^2\sigma^2(\bar{\omega}-\omega_0)^2] + exp[-4\pi^2\sigma^2(\bar{\omega}+\omega_0)^2]\}$$

(18)

In the asymmetric network with Gabor filters, similar derivation as the Eq. (16) is carried out. Then, the output of the sine-phase Gabor filter becomes the Eq. (19)

$$\sin\phi \cdot \{exp[-2\pi^2\sigma^2(\bar{\omega}-\omega_0)^2 - exp[-2\pi^2\sigma^2(\bar{\omega}+\omega_0)^2\},$$

(19)

while the output of the cosine-Gabor filter is the same as the Eq. (17). Thus, the output $R(\omega)$ of Eqs. (16) and (17) in the asymmetric networks becomes the Eq. (20)

$$R(\omega) = (\sin\phi + (1/8)) \cdot \{exp[-2\pi^2\sigma^2(\bar{\omega}-\omega_0)^2 - exp[-2\pi^2\sigma^2(\bar{\omega}+\omega_0)^2]\} \quad (20)$$

The Eq. (20) shows the two peak values are shown at the angular frequencies, $(\bar{\omega}-\omega_0)$ and $(\bar{\omega}+\omega_0)$. From the given frequency ω_0 of the Gabor filters, the peak value at $(\bar{\omega}-\omega_0)$, the angular frequency $\bar{\omega}$ of the input sinusoidal stimulus is obtained from $R(\omega)$. From the first peak value of $R(\omega)$, the phase information of the input sinusoidal stimulus, $\sin\phi$ is obtained from the Eq. (19). The energy model only derives the frequency information, $\bar{\omega}$, not its phase one, ϕ in the stimulus.

5   Response Relations for Complex Stimulus Changes

We assume the stimulus is moving as shown in Fig. 5, in which different shadowed stimulus on the B_1 and B_2 cells are moved to the arrowed direction. The stimulus on the B_1 cell is shown in the Eq. (21), while that on the B_2 cell is shown in the Eq. (22).

$$x'(t) = \alpha x(t) + \beta x''(t) \quad (21)$$

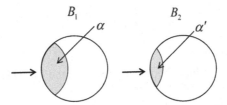

Fig. 5. Schematic diagram of the two stimulus for both cells

$$x'''(t) = \alpha' x(t) + \beta' x''(t) \quad (22)$$

This stimulus is shown in Fig. 5, in which the stimulus $x'(t)$ is inputted on the B_1 cell of the linear pathway in Fig. 1 and also $x'''(t)$ is inputted on the B_2 cell of the nonlinear pathway.

The final output of the asymmetric network by the two stimulus, become

$$\begin{aligned} y(t) &= y_1(t) + y_2(t) \\ &= \int h_1'(\tau)x'(t-\tau)d\tau + \iint h_1''(\tau_1)h_1''(\tau_2)x'''(t-\tau_1)x'''(t-\tau_2)d\tau_1 d\tau_2, \end{aligned} \quad (23)$$

where $y_1(t)$ is the output of the linear pathway, while $y_2(t)$ is that of the nonlinear pathway. The mean square of the error for the optimization of the network is given by

$$\xi = \int \{y(t) - y_1(t) - y_2(t)\}^2 dt \tag{24}$$

The conditions of the minimization of the Eq. (24) is given in the following equations.

$$\frac{\partial \xi}{\partial h_1'(t)} = 0, \; \frac{\partial \xi}{\partial h_1''(t)} = 0, \; \frac{\partial \xi}{\partial \alpha} = 0, \; \frac{\partial \xi}{\partial \alpha'} = 0 \tag{25}$$

From the first Equation in (25), the Eq. (26) is derived.

$$E[y(t)x'(t-\tau)] = h_1'(\tau)(\alpha^2 + k\beta^2)p \tag{26}$$

From the second Equation in (25), the Eq. (27) is derived.

$$E[(y(t) - C_0)x'''(t-\tau_1)x'''(t-\tau_2)] = 2h_1''(\tau_1)h_1''(\tau_2)(\alpha'^2 + k\beta'^2)p^2 \tag{27}$$

From the third Equation in (25), the Eq. (28) are derived.

$$E[y(t)x(t-\tau)] = h_1'(\tau)\alpha p$$
$$E[y(t)x''(t-\tau)] = h_1'(\tau)(\alpha\alpha' + k\beta\beta')p \tag{28}$$

The optimization Eq. (25), are derived to the following Eq. (29) using Wiener kernels [9, 10, 13].

$$\frac{C_{12}}{C_{11}}\sqrt{\frac{C_{21}}{C_{22}}}\frac{\{\alpha^2 + k(1-\alpha)^2\}}{\{\alpha'^2 + k(1-\alpha')^2\}} \tag{29}$$

The Eq. (29) shows the directional equation for the both directional stimulus. The Eq. (28) shows the relation of the left side Wiener kernels and the equations of stimulus parameters α and α' without impulse response functions as Gabor filters. To estimate α and α' roughly from the left side equations in the Eq. (28), the left side value is assumed to be a fraction (n/m), where n and m show the numerator and denominator, respectively. Then, the n and m values satisfy the Eq. (29) for the respective quadratic equations $\alpha^2 + k(1-\alpha)^2 = n$ and $\alpha'^2 + k(1-\alpha')^2 = m$.

$$\frac{k}{k+1} \leq n, \; m \leq k \tag{30}$$

As an example, $k = 2$ and $(n/m) = (0.7/1.5)$ are assumed, which satisfy the Eq. (30). Then, the quadratic equations of α and α' derives $\alpha = 0.56$ and $\alpha' = 0.20$, respectively.

Similarly, the Eq. (28) is derived in the Eq. (31).

$$(E[y(t)x(t-\tau)])^2/E[(y(t)-C_0)x(t-\tau_1)x(t-\tau_2)] = (G_s(\tau))^2\alpha^2/2G_c(\tau_1)G_c(\tau_2)\alpha'^2$$

$$(31)$$

From the Eqs. (29) and (31), the parameters α and α' are derived in the case of the Gabor filters $G_s(t)$ and $G_c(t)$ to be given. In the case of the conventional energy model with Gabor filters, the parameters α and α' are not derived.

6 Conclusion

The neural networks are analyzed to make clear functions of the biological asymmetric neural networks with nonlinearity. This kind of networks exits in the biological net-work as retina and brain cortex of V1 and MT areas. In this paper, the behavior of the asymmetrical network with nonlinearity, is analyzed to detect the directional stimulus from the point of the neural computation. For the motion detection, the asymmetrical network proposed here is compared with the conventional quadrature energy model with Gabor filters. It is shown that the asymmetrical network with Gabor filters works as like the conventional energy model. Further, the asymmetric networks detect the phase information for the moving stimulus, while the conventional symmetric networks do not find it. The responses characteristics to the complex stimulus can be computed in the asymmetrical network.

References

1. Reichard, W.: Autocorrelation, a principle for the evaluation of sensory information by the central nervous system. In: Rosenblith (ed.) Wiley, NY (1961)
2. Adelson, E.H., Bergen, J.R.: Spatiotemporal energy models for the perception of motion. J. Opt. Soc. Am. A **2**, 284–298 (1985)
3. Grzywacz, N.M., Yuille, A.L.: A model for the estimate of local image velocity by cells in the visual cortex. Proc. R. Soc. Lond. B **239**, 129–161 (1990)
4. Chubb, C., Sperling, G.: Drift-balanced random stimuli, a general basis for studying non-Fourier motion. J. Opt. Soc. Am. A **5**, 1986–2006 (1988)
5. Taub, E., Victor, J.D., Conte, M.: Nonlinear preprocessing in short-range motion. Vis. Res. **37**, 1459–1477 (1997)
6. Simonceli, E.P., Heeger, D.J.: A model of neuronal responses in visual area MT. Vis. Res. **38**, 743–761 (1996)
7. Heeger, D.J.: Models of Motion Perception, University of Pennsylvania, Department of Computer and Information Science, Technical report No. MS-CIS-87-91, September 1987
8. Heeger, D.J.: Normalization of cell responses in cat striate cortex. Vis. Neurosci. **9**, 181–197 (1992)
9. Marmarelis, P.Z., Marmarelis, V.Z.: Analysis of Physiological Systems – The White Noise Approach. Plenum Press, New York (1978)
10. Wiener, N.: Nonlinear Problems in Random Theory. The MIT Press, Cambridge (1966)
11. Sakuranaga, M., Naka, K.-I.: Signal transmission in the catfish retina. III. Transmission to type-C cell. J. Neurophysiol. **53**(2), 411–428 (1985)
12. Naka, K.-I., Sakai, H.M., Ishii, N.: Generation of transformation of second order nonlinearity in catfish retina. Ann. Biomed. Eng. **16**, 53–64 (1988)

13. Lee, Y.W., Schetzen, M.: Measurements of the Wiener kernels of a nonlinear by cross-correlation. Int. J. Control **2**, 237–254 (1965)
14. Ishii, N., Deguchi, T., Kawaguchi, M.: Neural computations by asymmetric networks with nonlinearities. In: Beliczynski, B., Dzielinski, A., Iwanowski, M., Ribeiro, B. (eds.) ICANNGA 2007. LNCS, vol. 4432, pp. 37–45. Springer, Heidelberg (2007). https://doi.org/10.1007/978-3-540-71629-7_5
15. Ishii, N., Deguchi, T., Kawaguchi, M., Sasaki, H.: Application of asymmetric networks to movement detection and generating independent subspaces. In: Boracchi, G., Iliadis, L., Jayne, C., Likas, A. (eds.) EANN 2017. CCIS, vol. 744, pp. 267–278. Springer, Cham (2017). https://doi.org/10.1007/978-3-319-65172-9_23

Author Index

Printed in the United States
By Bookmasters